The Ultimate Guide to
Raising Farm Animals

The Ultimate Guide to
Raising Farm Animals

A COMPLETE GUIDE TO RAISING CHICKENS, PIGS, COWS, AND MORE

Laura Childs, Michael and Audrey Levatino, Jennifer Megyesi,
Jessie Shiers, and Kate Rowinski

Skyhorse Publishing

Arrangement copyright © 2016 by Skyhorse Publishing

Text and images originally appeared in *The Joy of Keeping Chickens* by Jennifer Megyesi, *The Joy of Keeping Horses* by Jessie Shiers, *The Joy of Birding* by Kate Rowinski, *The Joy of Keeping Goats* by Laura Childs, *The Joy of Keeping Farm Animals* by Laura Childs, and *The Joy of Hobby Farming* by Michael and Audrey Levatino.

Skyhorse Publishing books may be purchased in bulk at special discounts for sales promotion, corporate gifts, fund-raising, or educational purposes. Special editions can also be created to specifications. For details, contact the Special Sales Department, Skyhorse Publishing, 307 West 36th Street, 11th Floor, New York, NY 10018 or info@skyhorsepublishing.com.

Skyhorse® and Skyhorse Publishing® are registered trademarks of Skyhorse Publishing, Inc.®, a Delaware corporation.

Visit our website at www.skyhorsepublishing.com.

10 9 8 7 6 5

Library of Congress Cataloging-in-Publication Data is available on file.

Cover Photographs: Thinkstock

Print ISBN: 978-1-63450-329-7
Ebook ISBN: 978-1-5107-0112-0

Printed in China

CONTENTS

Introduction

The Search for a Farm

We understand how you feel. We know why you're reading this book. You want to live the country life and enjoy the peace of mind it has to offer. You can imagine yourself enjoying breakfasts of eggs from your chickens and asparagus from your garden and honey from your own bees on your toast. In the winter, you see yourself snug and cozy by your wood-burning stove, warmed by wood from your own forest, that you gathered and split. You might even be wrapped in a knitted blanket made of wool from your llamas or alpacas. But you've got a long way to go before you get there; first things first, you need the farm itself.

Where will your farm be located? To determine this, you'll have to think about many more things than the sun and the soil. You'll have to imagine all the possibilities that a location might offer you. Is there sufficient land that's even enough—not too steep—to grow crops? How much sun exposure do all parts of the farm receive throughout the day? Are there fields of healthy grass that would support grazing animals or areas that you could clear for that purpose? Is the soil heavy clay or sandy grit? Are there any water features that might help irrigate a garden or help raise the water table near your farmhouse well? Is there enough forested land to manage a wood lot? As we talk about a sustainable hobby farm in this book, you'll see that there are many contributing factors to maintaining a healthy balance on a farm, including water, grass, animals, wood, and your own human spirit. Instead of thinking of your farm as a location, think of it as a living system.

Finding the right farm is both exciting and stressful, but it's not harder than any other house search, just different. You're making a lifestyle change, and because of that, it might feel like a much weightier undertaking. We recommend that you take extra time in your search, even if you have to rent for a while. Because it takes so many years to really work a farm into your own liking, making the correct choice initially is very important. And depending on how quickly you need to get settled, you're limited by the actual farms that are available at the time you're looking.

Generally, the further away from an urban area you get, the more affordable the property values become. But living further out means a longer trip to the grocery, the hardware store, and especially the markets where you might want to sell the fruits of your labor. Because you'll likely have an off-farm job, you'll want to consider the length of your daily commute as well. If you're an aspiring hobby farmer, some convenience to creature comforts is probably important to you. Today, you can find a location that takes advantage of rural solitude and urban comfort, if that's what you desire.

RIGHT: The view of the garden at Broadhead Mountain Farm from the house. Consider the proximity of the house to the garden, water sources, and outbuildings when purchasing a farm.

Tips for Farm Safety

✓ Wear protective eyewear and earplugs whenever operating machinery, even a lawn mower.

✓ Never operate machinery or a chain saw after having even one drink. Alcohol quickly impairs your judgment.

✓ Wear work gloves and closed-toe boots or shoes when doing any farm work.

✓ Avoid wearing loose clothing when operating machinery of any kind.

✓ Keep fire extinguishers in your kitchen and in your barn.

✓ Never operate heavy machinery or a chain saw without someone else nearby or at least at home that can call for help if you need it.

✓ Keep a first aid kit for humans and also one for your animals.

✓ Post all emergency numbers and contacts clearly on your refrigerator in case someone else needs to call for help for you or if something happens to your animals while you are away. And make sure to share your emergency information with your neighbors.

✓ Create proper fencing and regularly inspect it in order to keep your animals safe.

✓ Keep a calm and assertive demeanor when working around animals.

✓ Study the operator's manual of any machine before you use it.

✓ Never walk behind a tractor when the power take-off (the twisting knob on the back of a tractor that powers farm implements) is engaged. Clothing can easily become entangled or debris can be sent flying by cutting equipment.

✓ Always walk your field or lawn before mowing to look for hazards.

✓ Wear a protective face mask when spraying chemicals (even organic ones) or doing yard work to avoid allergy problems.

✓ Get to know your local plants, insects, snakes, and animals so you know which ones to avoid.

✓ Never put up wet hay in a barn. It can combust and is one of the leading causes of barn fires.

✓ Wear sunscreen.

✓ Keep children on a short and painful leash. Well, at least keep a close eye on them and teach them what they should avoid.

Buying tips for first-time animal hobby farmers:

- Buy locally and befriend the sellers so you might be able to call them for answers to questions you may have later.

- Read up on the specific animals you're buying and the traits to watch out for, like founder rings on the hooves of horses. But unless you plan to show or breed the animals, there's no need to be too picky outside of general health.

- Always buy two to begin with as farm animals do not do well living alone.

- Begin by buying gelded males. They are cheaper and you can learn to care for them and give yourself time to decide if you want to breed animals in the future. Buying a breeding pair instantly commits you to more than your original purchase.

- Visit the farm you're buying from at least once without actually buying. Don't buy on the spot. Take a look around, ask lots of questions, then go sleep on it for a few days.

- Stay away from livestock markets. They can smell a greenhorn coming from miles away and you'll never leave there with what you truly want or need.

- While we do advocate adopting from rescue organizations, don't do it initially. Rescue farm animals can have very serious physical and mental problems that may be beyond the expertise (or financial means) of a first-time farm animal owner.

- Start with the lowest-maintenance animals you can find, like llamas, donkeys, longhorn cattle, or chickens. It's much more fun and rewarding to grow into a menagerie at your own pace than to have them take over your life.

- Remember that even chickens can live twenty years or more. Horses, llamas, and other large grazing animals can live well into  their thirties. Keeping animals you aren't going to eat is a big commitment.

What Do All Animals Need?

Shelter—Most of the animals we recommend in this book need just a three-sided shelter with a roof to be happy. They'll only use it to get dry or stay out of the harshest weather. If you live in a very cold, wet, or hot climate or you're raising show animals, then you will need a barn to keep them during the coldest and hottest days of the year. (Chickens need a fully enclosed shelter to keep them safe from predators; more on chicken coops and tractors to follow.)

Food and water—The animals we recommend for the hobby farmer need very little in the way of extra food, as long as you give them enough pasture to graze. Animals are not healthy when they are overweight and you should not feed your farm animals in the same way you feed your dogs and cats. They should be left to forage mostly for themselves, except when that forage is not available. We only buy hay for the harshest months of the winter when there's not any green grass left on the ground. If you live somewhere that is very dry and has very little grass for much of the year, you should really think twice about having animals at all. Raising animals where they can't naturally forage for themselves is not sustainable. But animals can be made to create healthy pastures if you employ proper rotation techniques. The books of farmer Joel Salatin describe a sustainable grass-fed animal rotation system. Even in areas that do have good grass, we know of times when drought conditions have required farmers to travel several states away to find hay. The amount of on-farm grass and feed you can produce is a big factor in the cost of keeping animals.

Access to clean, fresh water at all times is very important. Some animals can share water sources. For instance, our donkeys and llamas share a tank and our house chickens, dogs, and cats all drink out of the same water bowls scattered around the yard. Make sure that the water doesn't freeze in the winter time. We usually carry buckets of hot water from the house to pour into the frozen bowls and tank to break up the ice in the winter. But electric water heaters are inexpensive and convenient too.

Basic veterinary care—Most large animals should be seen by a veterinarian once a year. For hardy animals like donkeys and longhorn cattle, you might get away with never needing a vet unless they have a problem. At a minimum, you should have your animal seen once by a vet to establish the relationship and to get their professional advice. Then you can decide for yourself how often you'd like your animals to be seen and evaluated. There's some debate as to how many vaccinations are needed for all the various farm animals. We tend to believe that most farm animals are over-vaccinated. The recommendations of your local agricultural extension agent are very much in line with their conventional recommendations for raising plants; they rely heavily on chemicals. But this will have to be a judgment call on your part. Some people suggest a West Nile vaccine for donkeys and llamas, but a few vets we've talked to have explained that donkeys seem to be less susceptible to this virus than horses. They suggest that the animal would have to have another pre-existing condition that weakened their immunity before it could be a problem. Tetanus shots are another issue. There's no clear evidence that indicates how often farm animals need to be vaccinated against tetanus. We've only had our donkeys vaccinated twice in the eight years we've owned them. But we do have our llamas vaccinated every year with whatever the vet recommends.

Companionship—The more attention you give your animals, the better. They like it. They become more tame and easier to handle. You can spot any problems early. And animals provide good, old-fashioned entertainment.

Preparing for Animals

Keeping in mind the basic needs of farm animals, you'll need to prepare for their arrival. If you have animals already, you'll want to make sure you have a separate area for the new

animals to go until they get settled. It's best to have a small pen next to the field where other animals are so that everyone can get acquainted over a fence for a day or two before intermingling. We use our dog pen (without the dogs in it, of course). A temporary fence of posts and ropes is okay, if that's all you have.

Even if you stick with the low-maintenance animals as we suggest and you have good grass, it's best to have some feed for the first few days. It's a good way to break the ice and to let the animals know that you are the provider. They can usually use a little more high-calorie and protein food if they're stressed from being moved to new place. Buy some good grain or sweet feed, but don't completely spoil them.

Walk your fields and fences before they arrive and replace any fencing that might be dangerous. If you've got barbed wire from having cows previously, you should really replace that if you are introducing horses, donkeys, or llamas. We spent a whole day gathering up old rusty barbed wire that had been left around our fields before we brought in our llamas. We then put them in our dog pen for the first night. But we didn't fully check the dog pen. There were some wires sticking out that were used to attach the woven wire to the posts. Within the first four hours, one of the llamas was spooked and caught his lip on the wire. It ripped it wide open and we had to call the vet out at 9 p.m. on a Sunday to stitch him up—on his very first day!

You should also make sure you have proper halters and leads for the animals before they arrive.

Can't We All Just Get Along? Yes!

Most of the animals you'll keep on your farm are natural enemies—dogs and cats, dogs and chickens, dogs and horses. Well, we guess it's really the dogs that are the problem. And there certainly are dogs that have that taste for chicken that you may not be able to cure. But we've found that if you slowly introduce your animals to each other, while supervising at all times, they will learn to get along. But the most important precursor to this is for you to establish your rightful spot as the pack leader. If your animals do not respect you, then they certainly won't have any reason to respect other animals in the pack.

Always keep animals separate initially when bringing them onto the farm. Keep them in a cage or pen that the other animals can approach and smell. Then introduce your dogs one by one to the new animals (you don't want them to get the pack mentality if they're all together) after a couple of days. Cradle the cat or chicken and get on the ground with the dog (best to do this inside, if it's a cat). Call the dog over and allow him to sniff. He will probably try to nip at the animal and that's when you give him a sharp rebuke to tell him what's appropriate. Then call him over again. When he gets that predator look in his eyes and lowers his head, sharply call his name to get his attention off the other animal. Keep eye contact with him and tell him what you expect. He won't understand the language, but he'll understand the tone. The most important part of this exchange is to get the dog's attention off the new animal and to direct it to your eyes so he knows you are completely focused on him. Spend time around all the animals together and make sure they know that you expect them to accept the newcomer into the pack. And whenever the dog seems to fixate on the other animals, sharply call his name and make eye contact with him again. If a dog will not ever completely calm down, our method of fixing this problem is to grab him by the scruff of the neck and look him directly in the eyes and assertively tell him, "No." Don't yell it. You should be very careful doing this, however. If the dog does not already respect you as the authority, he may bite you.

Sure, we've had our problems. We came home early one day to find our new rooster cowering behind a pile of wood with half his feathers ripped off. We probably didn't react the way the professionals tell you to. We picked up that rooster, ran after the dogs screaming obscenities at them, and locked them in their pen. They say that you can't correct a dog unless you catch him in the act. But our dogs knew what they had done and they never did it again. The rooster made a full recovery.

So stay around your animals when they are mixing until you are certain they have made nice. Never leave them alone together until you're absolutely sure everyone knows their boundaries. Sometimes it's just best to keep animals separate and that's what good fencing is for. Good fences make good neighbors, both for humans and animals.

Chhapter

Chickens

Chicken Basics

All chicken flock owners should familiarize themselves with basic chicken anatomy. It is useful when comparing other specimens of the same breed, and it is crucial to know what a normal chicken looks like in order to catch potential health problems before they are spread to the rest of your flock.

- The comb should be full and blemish-free in mature birds, indicating no frostbite, no underlying disease, and no scars from being overly aggressive toward other chickens.
- The eyes should be bright, clear, and alert.
- Nostrils should be free of discharge or crusty material. There should be no sounds emitted during normal respiration.
- The wattles should be rounded and undamaged—damage being another indication of fighting or exposure to extreme elements or disease. Check around the base of the wattles for parasites and their eggs. Lice will lay eggs in clusters that look like brittle clumps of whitish-colored debris.
- The plumage should be in good condition–bright, lustrous, sleek, and well-maintained.
- Check the vent for parasites as well; manure-caked feathers here may indicate disease or advanced age in birds.
- The breastbone should be straight, free of abscess, and well-feathered.
- The legs should be well-scaled and smooth. Inspect the toenails for debris. One indication of advanced age is the spur in the rooster; these are not well-formed until the second year of life.

Some Considerations Before You Get Started

Local and Federal Regulations

Local laws, such as zoning and other ordinances, may end your chicken enterprise before it takes flight. Some farming magazines have been advertising "Stealth Coops," mini-pens to conceal your poultry in more urban settings. Besides looking uncomfortable for the chickens (there doesn't seem to be much space for more than two or three chickens, nor are there sufficient locations for nesting, roosting, and foraging), the idea of concealing your chickens from your neighbors or from town and city officials seems like it will only lead to trouble.

The best way to avoid conflicts is to check with the county zoning board, your extension agent, or the state agricultural department. And after all the rules check out, consult your neighbors with your plan. Are they okay with the accompanying baggage associated with keeping chickens—noise, or an enticement for the dog to chase? Perhaps ruffled neighborhood feathers can be smoothed with the promise of fresh eggs and meat, or the reward of manure high in nitrogen for their garden.

" The fact that zoning in towns allows residents to raise a barking, crapping dog the size of a small elephant, but not four hens for a steady fresh egg supply shows just how lacking in common sense we have become as a society. The indictment against town chickens stemmed originally from roosters crowing at dawn and waking neighbors. The easy solution to that problem is to raise only hens or butcher the roosters before they mature enough to crow all the time. "
—GENE LOGSDON in *Chicken Tractor,* by Andy Lee, 1994

LEFT : Two Barred Plymouth Rock pullets, left, and a Partridge Plymouth Rock pullet roam the yard of Carol Steingress and Rick Schluntz. The couple have been raising hens for eggs in their small yard for the past 10 years.

Chicken Scratch

Do I need a rooster to have my hens lay eggs?
No. The rooster is needed if you want a fertile egg but isn't necessary for a hen to lay eggs.

Do hens lay more than one egg a day?
No. In fact, they will average less than an egg a day in many cases, and their productivity depends on many factors, including age, diet, weather conditions, daylight hours, and breed.

Are all roosters mean?
No, but some breeds have a proclivity toward mean roosters (see breed chart on page 14–15).

Do roosters only crow at sunrise?
No—they will crow when startled, when they are establishing hierarchy, or simply to announce their territorial presence.

Breed

Breeds of chickens are grouped together according to their size, shape, plumage, number of toes, color of their skin, etc. When mated together, individuals within a breed will produce chicks that share the same characteristics as their parents. Breed recognition can be different, depending on which poultry organization you consult. For example, the Marans is a French breed that is not recognized by the APA (American Poultry Association) because of its inconsistency in type (some Marans individuals in the United States have feathered legs, while others do not), but it is recognized as a distinct breed by the Poultry Club of Britain (PC). On the website www.feathersite.com, there are hundreds of breeds listed that occur worldwide. The APA only recognizes fifty-three large fowl and sixty-one breeds of bantams.

Class

Breeds are subdivided into classes. In large breeds, classes indicate their origin: American, Asiatic, English, Mediterranean, Continental, and Other (which includes Oriental). Bantam breeds are classified according to characteristics, like comb shape, or presence of feathering on the legs.

Variety

Varieties describe breeds based usually on plumage color, but also on comb style or feathering. For example, the Leghorn, in the Mediterranean class, has twelve varieties of large breeds.

Strain

A strain is a term that refers to a line of birds that have been bred for specific characteristics. In the show arena, strains are developed from a single breed for characteristics that are thought of as typical or "typy" and are considered superior by the owner and by fanciers of this same breed. There can be several different strains within the same breed. Commercial strains are often hybrids, sometimes having parents of different breeds. These strains are developed for superior production of either eggs or meat. A Cornish-Rock cross is a popular meat hybrid bred to grow heavy breast and thigh meat in a short amount of time. The Black Sex-Link is a laying hen that is crossed with a Barred Plymouth Rock and a Rhode Island Red that can be distinguished as a female from its plumage as a chick.

Foundation versus Composite Breeds

Poultry literature will often refer to chickens as being either foundation or composite breeds. A composite poultry breed is somewhat akin to a crossbred dog such as the Labradoodle, created using a purebred Labrador Retriever (a dog known for its easygoing personality) and the Standard Poodle (a dog known for its brains).

A foundation breed is a very old breed of chicken with distinct characteristics, such as the Dorking, with its five toes. This breed, along with Houdans and Asiatic breeds, was used to develop the composite breed in northern France called the Faverolle. The Faverolle was well adapted to battery cage production, where laying hens were confined in small cages for maximum egg production, and was a good source of winter eggs in the late 1800s.

Chicken Personalities

Behavioral traits among chickens vary widely. The American class contains breeds such as the Rhode Island Red, the Plymouth Rock, and the Jersey Giant that are generally docile, are cold tolerant, and can produce eggs and meat on a small scale. Most of the breeds in the Mediterranean class, like the Leghorn and Minorca, are flighty, smaller bodied, less cold tolerant, but more efficient in converting feed to egg production. The Dorking and Java are both excellent meat breeds, capable of foraging for themselves and requiring less grain, but they will mature more slowly than the Cornish Rock hybrids that are almost entirely dependent on being fed high-quality mixed rations. Dual-purpose breeds refer to those birds that produce enough good-quality meat and sufficient numbers of eggs that they could fit into small-scale operations for home use. It will be important to evaluate the traits of each breed to determine which would be most suitable for your enterprise. Several hatcheries offer assortments of chicks, so you can experiment in your first year or two. It might also be useful to visit neighboring flocks to discuss the pros and cons of the breeds with the flock owner.

"However reluctant those concerned with poultry may be to acknowledge the fact, it is not the less true that most old women who live in cottages know better how to rear chickens than any other persons; they are more successful, and it may be traced to the fact that they keep but few fowls, that these fowls are allowed to run freely in the house, to roll in the ashes, to approach the fire, and to pick up any crumbs or eat all the morsels they find on the ground, and are nursed with the greatest care and indulgence."

—SIMON SAUNDERS, from *Domestic Poultry: Being a Practical Treatise on the Preferable Breeds of Farm-Yard Poultry*, 1868

1. A Buff Orpington pullet roosts in the enclosed pen outside the coop owned by Cloë Milek and Karl Hanson on their 2/3-acre lot. The couple has been raising 20 hens for eggs for the past seven years. They have avoided keeping a crowing rooster to keep the peace with their neighbors.

2. A Black Star cross hen basks in the late winter light at at Luna Bleu Farm. The farm raises 100 hens for laying eggs and another 60 pullets are growing for their turn in the nesting box.

3. A Rhode Island Red hen, left, and a Golden Comet hen look at each other outside of the coop at Back Beyond Farm. Farmer Ray Williams prefers the Comets to the Rhode Island Reds, whom he has seen breaking eggs in the coop.

4. While Pistol the Old English Game bantam gives herself a dust bath, one of her chicks explores outside the coop belonging to Cloë Milek and Karl Hanson. Pistol hatched the chicks from fertile eggs Hanson got from a nearby farm.

3.

5. Rosie, a three-year-old Black Sex-Link hen, roams the backyard pen at the home of Geoff Hansen and Nicola Smith. When Hansen bought three pullets from a local farmer for their new coop, Rosie was included as a mentor to teach them the egg-laying ropes.

6. Danny, a Polish Crested rooster, is one of a menagerie of birds—including chickens—kept by the author.

7. A Cornish Rock cross meat bird rests at Back Beyond Farm in Tunbridge, Vermont. Farmer Ray Williams said customers prefer the variety's abundance of breast meat; the chickens weigh about seven pounds when packaged for sale.

8. With a puff of feathers around its neck and the dark eggs it lays, a Russian Orloff hen is a unique bird. The author was given the bird by someone who was moving away.

4.

5.

6.

How to Get Your Flock Started

Chicks can be hatched under a hen by saving eggs from your fertile flock (meaning there are roosters running freely with the hens). Fertile eggs can also be purchased and placed under a broody hen or in an incubator. Most commonly, chicks are purchased as day-olds that are mailed and delivered within twenty-four to forty-eight hours, but hatcheries usually require a minimum order to keep the chicks warm during shipping. Females that are close to beginning egg production, called started pullets, are probably most economical if you are interested in production as early as possible. In rural areas, local newspapers and agricultural magazines often advertise laying hens at a free or reduced cost. Show-quality birds are offered by breeders in fanciers' magazines, at poultry shows, or on the Web.

Before introducing new stock to an existing flock, the birds should be quarantined and given a clean bill of health from a veterinarian or a knowledgeable poultry producer. The National Poultry Improvement Plan (NPIP) maintains a directory of hatcheries and breeders that are enrolled in blood-testing programs for detection of several diseases that are contagious to poultry. The voluntary program was started in the 1930s as a means of eliminating pullorum disease from commercial poultry. Not all breeders are willing to wade through the red tape of bureaucracy, however, and you could always have your state extension agent do the testing on the birds that you are interested in purchasing.

Using an Incubator to Hatch Eggs

Incubators have several advantages over naturally hatching your chicks. First, you can time the hatch; you don't have to wait until a hen becomes broody. Second, you can hatch several more eggs at once with much less labor than it would take to hatch the eggs under broody hens that need individual attention, food, and water. Whether the broody hen will set on the eggs until they hatch is also a concern. Last, incubators are a great way to bring the joy of raising chicks into the home.

Descriptions exist of both the Chinese and the Egyptians hatching chicks by artificial means centuries before Christ's birth, in the time of Moses, by the ancient priests in the Temple of Isis. The Egyptians used sun-dried brick structures heated by fires, where the eggs were placed on grates in heated chambers. There were no thermometers in these chambers; the attendants who lived inside the incubators could sense the correct temperature to keep the fires.

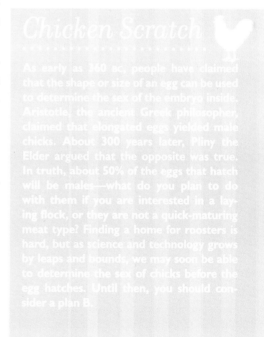

Chicken Scratch

As early as 360 BC, people have claimed that the shape or size of an egg can be used to determine the sex of the embryo inside. Aristotle, the ancient Greek philosopher, claimed that elongated eggs yielded male chicks. About 300 years later, Pliny the Elder argued that the opposite was true. In truth, about 50% of the eggs that hatch will be males—what do you plan to do with them if you are interested in a laying flock, or they are not a quick-maturing meat type? Finding a home for roosters is hard, but as science and technology grows by leaps and bounds, we may soon be able to determine the sex of chicks before the egg hatches. Until then, you should consider a plan B.

As many as 90,000 eggs could be incubated at a time, for 15,000,000 to 20,000,000 hatched yearly. Eggs were tested for fertility by placing them on the palm of the hand or against the face; if the egg was cold, it was discarded. In China, smaller ovens were used. The French used wine casks, packed with horse manure and circulated hot water, over which eggs were placed to artificially incubate; and in England, hot air was passed through flues and over the eggs to hatch chicks. But it wasn't until the early 1900s that incubators could boast hatching as high a percentage of chicks as could a setting hen. Modern incubators incorporate climate-controlled heat and humidity to simulate conditions that would naturally occur.

· · · · · · · · · Moving Chicks Outside · · · · · · · ·

Transitioning chicks from a pen that has been heated artificially, kept from wind, rain, and sun, and kept free of predators to pasture can be difficult. Before the move is made, you should:

- Wait until the chicks have developed their *scapular feathers*, the little band of feathers above their wings on their backs. The feathers look a little like the shoulder pads on football players. They act to insulate the chicks from sun, cold, or wet conditions. Typically, chicks develop these feathers at about three weeks of age. Chicks should not be moved outside if nighttime temperatures within the pen can't be regulated above 60 degrees Fahrenheit, otherwise, they'll spend most of their energy keeping warm rather than growing.
- Wait until weather patterns are stable for your move. Choose a day to transition the chicks when the next few days don't threaten steady rain, thunderstorms, heavy winds, or intense heat.
- Move the chicks well before sunset so they can grow accustomed to their new surroundings. Put familiar watering and feeding devices in their pens to encourage them to eat and drink immediately.
- Have at least three sides of their new pen enclosed, and provide a heat light on a timer for the first few days to provide them with heat and light at dusk through dawn. Red-colored bulbs will disrupt their sleep patterns the least and may attract fewer predators than the white bulbs. At night, the pens should be predator proof, either closed tightly, or surrounded by electrified poultry fence.
- Monitor the chicks closely through the first few hours in their new home. They should be scattered about, busy foraging, emitting few, if any distress sounds (high-pitched, constant peeping sounds instead of pleasant, cheerful notes of content).
- Don't mix different batches of chicks together at the time of the move. This can lead to bullying or distraction from eating and drinking.

RIGHT: Buff Silver meat birds huddle together in a chicken tractor—a portable coop. The coop allows for the birds to be on fresh grass and distributes their manure.

- Make sure that feeding and watering stations are sheltered so that the young birds are not forced out in downpours or extreme heat to forage. These can either be areas inside an enclosed pen that are covered, or individual range shelters that act to keep feeding stations dry. Changing the location of feeding and watering stations will also distribute areas littered with grain and wet ground. The birds will tend to defecate more near where they are fed and watered, so manure will be more widely distributed if the stations are moved daily.

- As the birds mature, your management of feeding areas and shelter will change. When the birds are small, their feeding stations may need to be changed less frequently. As they mature, you may find yourself moving feeding stations and pens every day where they seek out shelter during inclement weather and at night. You may need to rotate the electrified boundary fences to change out their yards. At Fat Rooster Farm, we start out each new batch of chicks that is introduced to pasture by providing an enclosed shelter surrounded by electrified poultry fencing. As the birds grow accustomed to their outside quarters, we remove the fencing, allowing them to mix with the other birds. There is still predator protection along the perimeter of the common yard, and the night shelters continue to be moved so that a manure pack doesn't build up where the birds roost during the night.

How to Feed Your Chickens

Proper feed storage will reduce the chance of spoilage, as well as discourage an abundance of skunks, rodents, birds, and insects that are attracted to the free lunch.

Again, mold is detrimental to chickens, and moisture and humidity will break down the nutrients in the grain, leaving it unpalatable and possibly toxic to your chickens. A plastic or galvanized trash can will store grain nicely; open bags can either be emptied into the cans or left open inside (be sure not to leave the little strings that come off the bags lying around; chickens will invariably find them and wrap them around their feet).

Commercially manufactured feeders such as troughs or canisters are designed to hang from the coop's ceiling to feed your flock. They hover just above the ground so that rodents have a harder time getting into them, and chickens are discouraged from naturally scattering the grain around with their feet. Certainly, chickens can be encouraged to forage on pasture by scattering the grain amongst the vegetation, but it will also increase the amount of waste.

Chickens like routines. They are easily put off when things change, like weather, length of daylight, or the time of day that they're fed. It's best to feed your chickens within an hour or two of the same time each day; they'll lay eggs more consistently if they're laying hens and put on weight more evenly if they're intended for meat. I like to feed the birds in the early morning, after daylight has begun, and again at least two hours before sunset, so the birds have enough time to forage and then return to their roost, fed and happy for the evening.

If you are attempting to reduce the amount of prepared food they consume to take advantage of forage or garden surplus, delay feeding their grain ration until midday (never withhold water from them at any time, even prior to slaughter). If they're confined to the coop and a small chicken yard, it might be better to feed them free-choice grain and supplement treats from the garden instead. I supply young stock from the time they are chicks with a summer squash or two from the garden to encourage them to eat more than just grain (they may not eat it right away, but after a few days, they usually associate the treats with good taste). Keep in mind that commercially prepared rations will put the most weight on your chickens the fastest and allow your laying pullets to mature the safest. You might save on grain bills by providing your chickens with other foodstuffs, but it will take longer to reach the end product. At Fat Rooster Farm, the trade-off is an acceptable one, as the goal is to decrease the inputs from off-farm while producing higher-quality, pasture-raised meat and eggs.

Laying chicks will require a different feeding regime than chicks that are destined for meat. Remember that a laying hen should develop slowly so that her body will be able to withstand the stresses associated with laying eggs for the rest of her life. A chick destined for consumption will be pushed to grow as quickly and efficiently as possible to achieve the best conversion of feed to meat production possible.

If you decide to raise the two together, you'll need to slaughter the meat birds first and then change the laying chicks' ration to one with less protein. Laying chicks should not be

ABOVE: A variety of Plymouth Rock cross and Barred Silver meat birds gather around the waterer in the pasture.

given food that is too high in protein because it will force them to mature too quickly. A hen forced into production at too early an age will consistently produce smaller eggs and be prone to prolapse.

A simple way to slow the growth of laying-hen chicks is to introduce a whole grain like oats or wheat into their ration after they reach ten weeks of age, about the time the meat chicks are slaughtered, and feed it combined with their grower ration until approximately twenty weeks of age. Then slowly change them over to layer feed (which contains higher calcium) for maintenance.

I prefer to raise my laying hens separate from my meat birds. The laying hens seem more precocious and tend to harass the slower, more docile meat birds, and it's easier to keep the feeding regime straight by having them separated. If you've decided on a dual-purpose breed (a non-hybrid that can be raised for both meat and egg production), it's easy enough to slaughter the roosters at ten to twelve weeks of age, then add a carbohydrate source such as oats to the remaining hens' ration to slow their growth and maturity.

Keeping Laying Hens

Laying hens in a larger enterprise employ similar husbandry practices to those used by the backyard poultry keeper. Differences between small- and large-scale flocks might be in the breed of hen chosen, the feed they are given, and their living quarters.

Which laying hen breed should I choose? Each breed of chicken has a different rate of lay based on body type, ability to perform under certain environmental conditions,

Whitey, a White Plymouth Rock, wanders amongst the bushes at Carol Steingress's and Rick Schluntz's home. The couple have been keeping 5 to 6 laying hens in their small in-town yard for the past 10 years. "I think they're so interesting and curious," Steingress said. "We're hooked."

ability to convert feed efficiently to egg production, and other factors. Generally speaking, commercially developed hybrids produce more eggs. Most of these have been developed using the Leghorn, a breed in the Mediterranean class of chickens. They are flighty, small-framed birds that convert feed more efficiently to eggs than other breeds. Leghorns are not the best to choose if a dual-purpose bird is what you're after. The carcass of a Leghorn can't compare to that of a Rhode Island Red or a Plymouth Rock when it comes time to cull them from the laying flock due to poor laying performance.

Egg color may also factor into your decision. It took us years for our customers to accept our white eggs. To this day, people still ask us if they are nutritionally equal to brown eggs. Now, we mix our cartons, so that each one contains white, brown, and green eggs.

Generally, the color of a chicken's earlobe indicates the color of egg it will lay; white earlobed hens lay white eggs, while red earlobes indicate a brown egg layer. This rule runs into trouble when you have Silkies, whose earlobes are a robin's-egg blue.

How Eggs Are Laid

When a female chick is hatched, she is equipped with two ovaries. The development of the right ovary

Chicken Scratch

Are white eggs as nutritious as brown eggs?

The poultry demographic is clearly segregated. In New England, there are more brown eggs in the supermarkets, while in the Grain Belt and the South, white egg layers are more popular. The New England adage, "Brown eggs are local eggs, and local eggs are fresh" was true when the white eggs bought at the store were shipped from farther away. With modern-day transportation, this no longer applies, and on your farmstead all eggs, be they brown, white, or green, will share the same quality. There is no difference in the cholesterol content in brown versus white eggs, but free-range chickens (i.e., pastured, and allowed to forage at will) will produce eggs higher in omega-3 fatty acids than confined breeds. Confined birds can, however, be given feed supplements to increase omega-3s in their eggs.

stops in order to accommodate space for development of her eggs. Hens are hatched with the capacity to produce more than four thousand eggs in a lifetime, but even the hardiest of hens, like the Leghorn, will lay just under three hundred eggs in her first year of life, and then just over two hundred eggs the year after. A pullet's first eggs will start out small, but by thirty weeks of age, they will reach their normal size. As a hen ages, her eggs will grow larger. Hens can lay eggs for more than a decade, but most are culled from commercial laying flocks after eleven to twenty-four months, when production decreases substantially.

Where to Find Chickens and What Kind to Buy

Most people order their chicks through the mail from a reputable hatchery. The balls of fluff show up in a box and you put them under a heat lamp and keep them safe until they are ready to be let out into the coop. Most of the hatcheries are very good at preventing the spread of diseases and they will even inoculate the chicks against common diseases (although you may not be able to call them organic, depending on what they are inoculated with and when). But you're still contributing to factory farming if you go down this path. Hatcheries are huge operations that churn out chicks by the tens of thousands.

We suggest that you find a local farmer who sells eggs at the farmer's market. Ask around who has the best eggs consistently and who raises their chickens humanely. You may have to wait a little while until they are hatching chicks. When you find a farmer willing to sell to you, visit the farm and see how the chickens are cared for. Chickens with

A note about antibiotics and beak trimming (or docking):

Rob Rahm of Forrest Green Farm (Louisa, Virginia) is the most productive small egg producer we know. He maintains two flocks of about a hundred birds in mobile chicken coops. He also added a greenhouse roosting area that is pulled behind the main coop. This coop stays warm for the chickens and he was able to produce almost as many eggs through the winter as he has in the summer, proving that you need not use artificial lighting and confinement to keep your chickens laying all year (except in very northern climates of the country). In order to maintain the optimum health of the chickens to keep them so productive, Rob warns against ordering chicks that have been given antibiotics or have had their beaks trimmed. This is the standard procedure at hatcheries and you must make a special request that their beaks not be trimmed and they not be given antibiotics when you order. The argument (which makes a lot of sense to us) is that if you give chicks antibiotics very early in their development, then you weaken their immune system. When you get them home and stop giving them the antibiotics, they haven't built up a natural immunity to dangers in the environment and thus are more susceptible to becoming sick. Chickens with trimmed beaks (this is usually done to prevent them from eating each other's eggs) cannot forage for themselves properly and cannot catch bugs as well, also affecting their health and egg quality. You're just going to end up spending more on feed if you have chickens with trimmed beaks. We've found that chickens that are given enough room to roam and forage don't eat eggs and rarely get sick. So don't give the antibiotics, don't trim their beaks, and make sure not to overcrowd them.

a lot of room to roam and clean living conditions, without trimmed beaks or wings, will produce healthy, disease-resistant chicks. Or you may get lucky, as we did recently, and find a farmer who wants to sell you her entire chicken outfit as a turn-key operation. That way you've not contributed to the breeding of yet more farm animals. It's always best to have patience in finding animals and they usually end up finding you.

The other consideration is which breed to buy. There are literally hundreds to choose from. But if you'd like to keep your chickens around for the long run as layers, then you want a laying breed and not one raised for meat.

Chicken Advice for the Hobby Farmer

- You don't need a rooster. Hens will lay eggs with or without a rooster. So unless you want to breed your chickens, roosters are more trouble than they're worth (we've both got scars from rooster spurs to prove it). They are handsome, though. And there's something to be said about the comfort of that alarm clock each morning (roosters crow all day, not only in the morning). Never turn your back on an aggressive rooster. They will attack you and you will bleed as a result. Roosters, by our observation, do not feel pain when in defensive or offensive mode and have absolutely no fear when they become aggressive. Another reason to do without.

- Never let your chickens hatch their eggs unless you are absolutely sure you want to deal with the consequences: You will end up with a bunch of roosters that you'll need to either be prepared to eat or find someone else that wants a rooster, because roosters do not live well together (there are rare occasions when they do) and will fight each other to the death eventually. The flock we inherited recently has about eight cockerels beginning to fight each other. We've found another farmer who's interested in making them into soup and that's where they will end up. We hate to do this, but since we inherited these animals, we have to deal with them properly. It's either soup or a bloody fight to the death until the last one is standing.

- Chickens will eat anything. Keep hazardous materials away from them. We've witnessed our chickens eating paint off the side of the house (thank goodness it wasn't lead paint). We even witnessed a chicken steal a dead mouse from one of our cats and swallow it whole. A farmer friend of ours, who hunts deer in the winter, feeds the leftover carcasses to his chickens. He says it looks like someone's sandblasted a skeleton when they're done.

- If you want top-quality eggs, your chickens need a *lot* of room to roam. Our house flock (never more than about a dozen) has two and a half acres of fenced area to graze. So the eggs of our house chickens are richer in color and have firmer yolks than even our grass-fed chicken-tractor birds. Chickens typically lay an egg a day. If you want eggs only for your family, two or three chickens could be plenty. Eggs can last six weeks or more in the refrigerator.

- You can expect about 75 to 80 percent of your chickens to produce one egg a day in their peak during their first and second years. This drops to about 50 percent in the winter and can go as low as 20 percent when they molt. When chickens molt, they lose and replace their feathers and much of their energy is spent on feather production. You can help offset this, the same effect that cold weather has on them, by feeding them higher-protein food during their molt.

- In most cases, the color of chickens' ear lobes (the round spot straight back from each eye) will determine egg color. Brown or red lobe = brown egg. White lobe = white egg. But this isn't a hard and fast rule, especially with green eggs.

- If you don't want your chickens scratching around your yard or you can't keep your dogs from killing them, then go with a chicken tractor or mobile coop (see photos). This is by far the best setup if you want to get a steady stream of eggs, keep the chickens contained but not confined, and renovate the grass in your fields.

- Chickens get sick and chickens die. Sometimes it can be very messy and sad. Be prepared for it. There are a million diseases and health issues to which chickens are susceptible. You'll tie yourself in knots trying to diagnose them from books (a near impossibility, we've found). And there are few vets even willing to look at a chicken. The only one in our area charges $80 just for the initial exam. So if you have chickens, be prepared to put them down to keep them from suffering. We describe how to do that below.

Also, there are some breeds that are more aggressive, like Rhode Island Reds, that have terrific eggs but will do real damage to your flower beds, and the roosters can be quite ornery. We've found that almost all white chickens and roosters are docile and easier to put up with.

Shelter

If your chickens are free-roaming, like our house chickens, then they don't need much of a coop except to keep them dry and free from the wind and cold. Since we have five-foot woven wire fencing enclosing two and a half acres around our house with the dogs as guardians, then we only have a small coop with a couple of roosts and a couple of laying boxes for our house flock. The chickens only spend the night in there or go in when they need to lay an egg. A house flock like ours is

LEFT: Temporary housing for the pullets we inherited from another farmer. This coop can be pulled around the yard with our lawn tractor.

the easiest and lowest-maintenance way to keep chickens. If you're keeping your chickens confined to a smaller area, you'll need a bigger coop that offers them at least two feet of space (more is suggested) for each when they are all in the coop.

Another reason to give chickens as much room to roam as possible is that you'll have less to clean up after. A coop that has chickens in it all the time needs to be cleaned once a week at least. And chickens confined to a small yard will quickly turn it to dirt and mud (albeit nutritious dirt and mud). Our house chickens' coop typically only needs cleaning once a month. And by the time we do, most of the manure has begun to compost and it's easier and less smelly to deal with.

At night, chickens like to roost and bunch together. Chickens are easily trainable to go into a coop at night. If they don't do it on their own, wait for them to start to bed down for the night, catch them, and put them in the coop. Usually it only takes a couple of nights for them to figure out where they should go to sleep. Once they're in at night, shut them in and latch the door. Predators are very good at finding ways into coops. In the morning, open the door to let them out.

Chicken Tractors and Mobile Coops

Chicken tractors (which are movable chicken coops) are the way to go if you want more birds than a regular yard will handle, if you want to produce eggs for sale, or if you're interested in keeping your pastures healthy for other grazing animals. The coop is moved every few days with a tractor, truck, or lawn tractor and chain in order to give the chickens fresh grass and to spread their manure over the fields as a natural fertilizer rich in nitrogen. You can leave them on one spot longer to really give an area a natural boost. Parking a chicken tractor over your garden in the winter is a great way to boost your soil. Just be careful you don't overdo it and end up giving your garden a nitrogen imbalance.

We only advocate chicken tractors that have an electric fence enclosure to give chickens lots of room to roam and forage. The popular chicken tractors that are low to the ground and fully enclosed are just glorified cages. Sure, the chickens get fresh grass every few days. But chickens like to roam, peck around, flap their wings, and run after bugs. They like to roost above the ground at night. And it shows in the color and richness of the eggs. The more confined chickens are, the more you will spend on off-farm food to feed them. We have a chicken tractor we bought from a farm that was shutting down. It's the kind described

RIGHT: Our chicken tractor surrounded by an electric fence. The chickens will be moved around the garden to fertilize and eat harmful insects.

ABOVE: The enclosed coop at Broadhead Mountain Farm has a specially designed roost to catch chicken manure for use in the garden.

above, but we use an eighty-foot electric fence around it to give them room to roam. But it's difficult to move with the chickens in the coop, and then it becomes a trick to get them inside the fence, move the coop, then get the fence back up without them escaping. When we've tried to move the coop with the chickens inside, they become frightened and have been caught under the coop as it's being moved (never being seriously injured though). So it's not the most convenient way to keep chickens, although it is very good for fertilizing the ground. You should move them every few days, depending on how fast they are eating the grass and how much grass you have to begin with.

A mobile chicken coop on wheels or that's raised above the ground is, in our opinion, the best method. If you're using an electric fence (solar electric is quite convenient), just wait for the chickens to bed down in the early evening, close them in the coop where they are roosting comfortably, roll up the fence (make sure to turn it off first), move the coop to fresh grass, and set the fence back up. There's no drama of frightened chickens running this way and that. This can even be adopted as a smaller version for your yard without the electric fencing, as long as you have a door to shut them in or a dog to keep guard at night.

Enclosed Coops with Chicken Yards

This is the least economical option for raising chickens and produces eggs only as good as the grain you feed the chickens. But if you live in an area that has little grass for forage or somewhere that prevents you from rotating your flock or letting them roam free (like the mountains where there might be lots of predators), then this might be your only option. But be careful not to overcrowd them or they will resort to egg eating and cannibalism.

Food and Water

Chickens will eat anything. But *anything* gets into the eggs you eat. So be picky about what you feed them. The regular chicken feed and scratch at the feed store has pesticides and other chemicals you probably want to keep

ABOVE: Water and food are placed near the door for easy access.

out of your eggs. Our local feed store now sells organic grain, but it's quite expensive. After some time looking, we've now found a local chicken feed producer that sells non-GMO food without chemicals. But it's not certified organic. Our customers at the market don't seem to care about an organic label; they only want our word that we're not giving them any chemicals or antibiotics in their food. If your chickens have as much room to roam as our house chickens, then you don't even need to feed them except in the winter when there's no green grass. Given the choice, chickens eat mostly grass and bugs (more bugs result in deeper colored yolks). So in the winter, you may need to supplement their grazing. The chickens in our chicken tractor eat more food than our house flock, but we try to keep it to a minimum and force them to eat the grass instead. There's no formula if you're keeping mostly grass-fed chickens. You just need to pay attention to the amount of grass and grain being eaten and the production of eggs.

We combine equal parts organic chicken scratch with the non-GMO layer feed and add oyster shells (the bag will tell you the proper ratio) for calcium and grit to help them digest their food. You can also just leave out oyster shell or grit in a bowl for them to feed on when necessary. If you only have a house flock, just throw the food on the ground and let them forage. Only throw enough

for them to feed for about five minutes and you can do it in the morning and the evening. If you're keeping them in a smaller area or in a chicken tractor, then you'll need a feeder in the coop for them. If they have plenty of grass, they shouldn't eat more than a quarter pound of feed a day per chicken. We've found that chickens also like table scraps, breadcrumbs, the crushed chips at the bottom of the bag, and especially grapes. They love grapes. Meat has pathogens, so don't feed it to chickens. If you've got a chicken tractor operation, then you'll need a little trial and error to figure out how much feed to give them. It all depends on the size of the flock and the size of the area they are foraging. You want to encourage them to forage, so don't let them fool you into thinking they'll only eat grain.

LEFT: A solar electric fence makes this mobile chicken coop easy to move anywhere on the farm.

Chickens need water at all times. Dehydration will kill a chicken very quickly. A droopy, dull comb is a classic sign of dehydration. For our house flock, we keep dog bowls filled around the yard and all the animals use them. For our chicken tractor, we have one large watering can that we fill each morning and hang from a chain connected to the top of the coop. You might consider a water heater if you live in a place where it

ABOVE: A chicken house built on a hay trailer is easy to move and sturdy enough to carry large barrels of feed and water.

stays really cold for long periods. During the coldest part of our winter, we just fill a couple of five-gallon buckets with really hot water and go around and pour it into all the bowls. This typically unfreezes them and cools the hot water off enough for the animals to drink. Chicken tractors and coops need both grain and water feeders at all times.

Can You Profit from Your Chickens' Eggs?

Let's begin with a quick calculation of the costs. Perhaps you got lucky and found a used chicken tractor or mobile coop already built and you got a great deal. But you had to buy an electric fence with a solar charger, by far the easiest to use for a pasture-fed operation. You bought fifty chickens close to laying age in the spring. Add in some feeders, water troughs, and various food scoops. You'll need about three bags of supplemental feed each week and some oyster shell or grit. You'll need several sealed plastic containers to store feed. Egg cartons are twenty-five cents apiece on a good day, although you might find that customers bring you used cartons once you get established. So using very conservative estimates:

The chickens will start laying gradually and will finally reach their peak at the beginning of the summer. Say you have a really productive flock and you get forty-eight eggs a day out of fifty hens (this is a rosy scenario; 75 percent egg production is more typical). That's

50 chickens near laying age ($6 each) = $300
Chicken tractor (used) = $250 (new ones are $800 to $1,000)
Electric fence = $175
Solar electric charger = $185
Feeders (2 @ $18), water troughs (2 @ $20), and metal scoops ($24) = $100
Plastic feed containers (2 @ $35) = $70
Enough of the cheapest feed and oyster shell you can buy to last the summer (five months)
     = $700 (this can be much higher if you go with organic)
Total = $1,780

four dozen a day, seven days a week (chickens don't rest on Sundays). That comes to twenty-eight dozen a week. The average price at our market is about $3.50 a dozen, although some people get as high as $6. That's about $98 week in sales. So it takes roughly eighteen weeks to break even, if you're lucky. Chickens slow down production in the winter when it's cold and when they go into molt. Also, you might have a harder time reaching retail customers in the winter and you might have to sell them cheaper to a reseller. If for the rest of the thirty-four weeks out of the year, you average 20 dozen, that's 680 more dozen you have to sell. That would be $2,380 in profit, assuming you have a steady direct-to-customer base and you don't have to sell wholesale, which would cut that amount in half. So, you stand to make a maximum of $2,380 in your first year, raising fifty grass-fed chickens and hustling to develop a loyal customer base you can sell direct to year round. They never keep up the same rate of laying after their second year and will steadily decline, but you also won't have the startup costs each year, so after about three years you might double your profit.

If you can do this, you'd be the most successful chicken egg hobby farmer we've met. But there are many variables in raising chickens, especially if they come down with a communicable disease. One farmer we know had his flock come down with one of the many diseases that will make chickens quit laying for good. He had to put down ninety birds himself and he didn't feel right about eating them or selling them for meat since they were diseased (most of the producers you find in the supermarket aren't so discerning). A heartbreaking task, both personally and economically.

Tom Martin of Poindexter Farm in Virginia buys chickens just as they're ready to lay and sells them at the end of each farmer's market season to friends and neighbors that want laying hens so he doesn't have to pay to feed them through the winter when they're less productive. In this way, he's a bit more profitable than most small chicken farmers. But as you can see, it's very hard to get rich selling eggs. And raising chickens for meat is not that different than raising them for eggs. Our economic formula would only need slight tweaking for meat birds and your potential profit would be about the same. Your labor would certainly go up, though, as you'd need to kill and dress them. Unless you're going to grow exponentially and make the jump to becoming a factory chicken farmer, it's not the way to riches.

Money isn't the only benefit you get from raising chickens, however. Knowing that your eggs come from chickens that eat all-natural foods and aren't given any antibiotics or chemical-laden feed is worth a good deal to most people. Chickens also create valuable fertilizer for your garden and renovate your fields without chemicals. They eat lots of pests as well. They are one piece in a big puzzle of diversity on your farm. So like all the hobby farming ventures we promote in this book, don't do it to get rich. Do it to enrich your life and your farm first and if there's a small profit from it down the line, then all the better.

Apple Cider Vinegar

Apple cider vinegar is an all-natural supplement that you can add to your chickens' water in a quarter-cup per gallon ratio. It has vitamins, minerals, and trace elements. It lowers the pH in chickens' stomachs to help with digestion. It's also an antiseptic, which helps kill germs and boost a bird's immunity. But you'll want to give your chickens the raw, unfiltered kind so that they get all the benefits. Most health-food stores or Whole Foods carry it. And you might find it in your local feed store or co-op. Also, only use plastic water containers, as the vinegar will corrode metal. Some farmers give it as a regular supplement to ward off disease. Others wait until they see a problem with their chickens before using it. The science isn't in on which option is best, so a little trial and error is the best course of action.

When Something Goes Wrong

Chickens are tough creatures physically, but they are very susceptible to disease. And it's very difficult to diagnose a chicken unless you've had a ton of experience and worked with someone who knows all about chicken problems. Books will drive you mad like online medical sites do. You'll find yourself believing all your chickens have avian flu by the time you put the book down. Chickens get parasites, frostbite, bacterial problems, and coccidiosis (coxy), which is probably the most common chicken disease. They can also have egg binding, which is when an egg gets stuck coming out. This is one of the easier problems to fix and involves a little Vaseline and a rubber glove.

You'll know when there's a problem. A sick chicken will stay away from the rest of the flock, lower its tail, puff up its feathers, and not move or graze much. If you spot a problem, isolate the chicken from the others. You do need a separate cage for this purpose. A dog crate works well. Make sure the chicken has food and water, and put electrolytes in the water. That's about all you can do unless you find someone who knows something about chickens that can help. Most of the time, the chicken will perk back up soon.

But sometimes chickens get sick and don't get better. They get weak with diarrhea and become dehydrated. Sometimes you'll just find them dead. Other times, they have a serious

When a Chicken Meets Its End

We've had to put other chickens down since then too. This is the easiest way to do it that seems the least traumatic for chicken and human:

- First and foremost, be calm and deliberate. Suck it up. It's your responsibility to keep the animal's suffering to a bare minimum. Prolonging that suffering because of your own emotional frailty is unacceptable.

- Hold the chicken upside down for a couple of minutes by its legs. This allows the blood to rush to its head and makes it calm.
- With its comb facing away from you, pin its head on the ground with a broomstick or something similar.
- The stick should be right behind the head and you hold the stick down on either side of the chicken's pinned head with your feet.
- Now, with the chicken's head pinned between your legs, pull up solidly on the chicken's feet so that its head and neck stretch. This separates the head from the spine.
- Hold that pressure (not too much or you'll have a headless chicken, but that works too) until the chicken stops flapping its wings and dies. This is a nervous system response and doesn't seem to create any suffering.
- Dig a deep hole where your dogs can't get to it, say a few words of praise, and bury it.

problem and it's obvious they are suffering. We had two roosters for a time. They got along well enough until one of them started showing signs of age. The younger one would periodically beat up the older one (you've probably heard of the pecking order) and we'd have to separate them. One day, the younger rooster really let the old guy have it and wounded him to the point that he couldn't walk. So we had to put him down.

Goats

WHAT'S NOT TO LOVE about goats? They will feed you, clean up the overgrown mess in the fields, and take long hikes with you while carrying your supplies. They will make you laugh when you're sad, provide extra income for even the smallest farm, carry you to town and back in a little cart, and perhaps best of all, they will gaze upon your face with earnest adoration.

Most of us though, are excited about the prospect of keeping goats for their deliciously sweet milk and low-fat, nutritious meat.

Goats' milk is consumed in larger quantities by more people, than cows' milk. A staggering 65 percent or more of the world's population choose goat over cow's milk for reasons other than availability or economics. Perhaps even more surprising, over 60 percent of all red meat consumed worldwide is goat meat. Goat meat has substantially less than half the fat of chicken, beef, pork, or lamb.

To keep, the goat is both a pleasure and all-round multitasker. Traits that are well appreciated as we seek the means to cut back on stress and find joint purpose for every facet of our lives.

Goats provide many blessings for minimal cost and care and can easily be kept on just a few acres of land. They are a livestock staple in all of the old-world European countries and are finally gaining the attention they deserve from North American farmers.

The goat has been called the poor man's cow for far too many years. Although their service may initially fit the bill for the frugally minded, there is little chance of counting them as a lesser animal once you start tabulating their virtues. You may have one main reason for your interest in goats today, but once they are in your keep, you can't help but notice how wonderfully suited they are to a farm or family's needs. In time you will have forgotten why you bought that first goat, being left only to wonder how you ever got along without them.

Goats have been a barnyard staple in European countries for centuries, but we are only just now realizing their multiple uses and gifts.

Your appreciation of the goat won't stop at the barn door though. Every goat you pour your time into will also be working their goat-given charms on you. I have not met one goat keeper, even those with large herds who hasn't been completely enamored with or emotionally connected to a favorite *Capra aegagrus hircus* (Latin: goat).

As you reap the gifts they repeatedly bestow, you'll wonder why you waited so long to discover the joys of keeping goats.

Basic Goat Anatomy

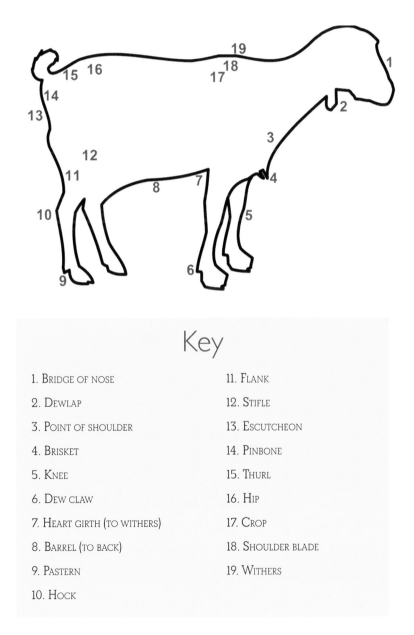

Key

1. BRIDGE OF NOSE

2. DEWLAP

3. POINT OF SHOULDER

4. BRISKET

5. KNEE

6. DEW CLAW

7. HEART GIRTH (TO WITHERS)

8. BARREL (TO BACK)

9. PASTERN

10. HOCK

11. FLANK

12. STIFLE

13. ESCUTCHEON

14. PINBONE

15. THURL

16. HIP

17. CROP

18. SHOULDER BLADE

19. WITHERS

Body Condition Scoring Assessment

Once you have a few goats in your barn, you will be able to tell a lot about their mood and well-being at just a glance. Healthy goats have shiny coats and are curious by nature. They hold their head, ears, and tails up and have a firm, confident stance.

We all know that we can't judge a book by its cover though. To truly assess a goat's condition, whether you're considering a purchase or adjusting feed quantities, you'll want to perform a body condition scoring assessment.

Body condition scoring is a standardizing system developed for all animal types. Goats are no exception. Use the chart below to perform a hands-on assessment for goats.

BODY CONDITION SCORING ASSESSMENT

Score		Spinous Process	Rib Cage	Loin Eye
BCS1	Very thin	Easy to see and feel the sharp point of bones can be felt.	Well pronounced, easy-to-feel top of, and slightly under, each rib.	No fat covering.
BCS2	Thin	Easy to feel but smooth.	Smooth and slightly rounded. Slight pressure is required to feel individual ribs.	Smooth with a slight, even fat cover.
BCS3	Good Condition	Smooth and rounded.	Smooth and even to a gentle touch.	Smooth, even fat cover.
BCS4	Fat	Can feel with firm pressure, but no points can be felt.	Individual ribs can not be felt, but you can still feel a slight indent between them.	Thick fat cover.
BCS5	Obese	Smooth; no individual vertebra can be felt.	Smooth to the touch. No individual ribs can be felt. No separation of ribs felt.	Thick fat covering. The fat may feel lumpy.

The spinous process is in reference to the vertebra of the spine. When you first touch the goat for assessment, run your fingers down the spine from the shoulders to the tail head and assess the feel of the bones and padding between hide and bone.

Next, touch and assess the rib cage on both left and right sides, from the top of the back to the under side of the goat.

Finally, stand behind the goat and position both of your thumbs on the spine just behind the start of the rib area. Curl your fingers down and back as though you were going to pick the goat up using just your hands on the muscle there. You now have the loin area cupped in your hands. This will be the most difficult area to assess until you have practiced on goats of differing condition.

More advanced assessments can be performed, but these are difficult to learn without assessing many different goats. For the best visual representation of body condition scoring,

visit the Langston University research website at http://www.luresext.edu/goats/research/
bcs.html.

Goat Speak—A Few Terms You Need to Know Now

Throughout this book and in your journey of discovering more about goats, you'll run across new terminology. Here are the most common.

- Dam—A goat's mother.
- Doe—A female goat. Young, unbred does are called doelings.
- Buck—A male goat. Young bucks are called bucklings.
- Brood Doe—A doe kept for breeding purposes, usually possessing excellent genetic traits or ancestry.

BELOW: Side by side, it isn't that hard to tell which one is the goat and which one is the sheep, but sometimes it isn't all that easy.

- Chevon—Meat from mature goats.
- Cabrito—Meat from goat kids.
- Cud—The regurgitated food of a ruminant.
- Herd—Two or more goats.
- Kid—A newborn goat. They are called kids until they reach a year of age.
- Ruminant—Any animal with four stomachs that chews and regurgitates their food.
- Sire—A goat's father.
- Teat—The two protrusions from the udder that milk flows through.
- Udder—The organ of a doe that produces milk.
- Wether—A castrated male goat.
- Yearling—A goat between one and two years of age.

Average Yield for Full-Size Registered Breeds

- Dairy—The average dairy doe supplies 900 quarts of milk per year; 2–4 quarts per day.
- Meat—The average buck kid provides 25 to 40 pounds of meat. Stockier meat breeds such as the Boer provide twice as much.
- Fiber—Adult Angoras supply 10 to 15 pounds of mohair per year. Adult cashmere producing goats might supply one-third of a pound per year.
- All—Supply approximately 1 pound of garden enhancing manure per day and excel at weed and brush removal.

Your Goat-Keeping Strategy

Before we discuss the various breeds, let's first explore a few farm strategies. Defining your strategy is as important, if not more important, than choosing the breed to suit your needs.

First of all, you should have at least two goats. Goats are herd animals; and they truly are happiest when they have one of their own kind to mingle with.

In most strategies you'll want two does, two wethers, or one of each. Keeping a buck or a breeding pair never fares well for goats. Full (unneutered) bucks produce a strong-tasting meat that is uncomplimentary for the North American palate. If a buck is housed with a doe, their presence will taint the delicate taste of milk and scent the hair of the doe. More importantly, if you keep a buck with a doe in pasture, you will lose control over all aspects of breeding.

At the very least I ask that you have at least one barn buddy to keep your goat company. Goats make great companion animals for a horse or cow. They also pair up nicely with sheep,

Goats are happiest in a herd of their own kind, but they also make good companion animals for horses.

but since their mineral requirements differ, sheep should never be allowed access to the goat's mineral block.

At any rate, if one goat is a charming addition to your farm, two will be twice the fun. So let's explore the options of keeping at least two goats and how it might benefit you. We need to first consider the output of each goat and how you'll put that output to use for your benefit. These are small farm or backyard farming strategies; if you plan on having a large herd, you likely have other ideas on management and production.

Goats Earn Their Keep

- Dairy—Aside from the obvious option of selling your dairy goats' offspring once or twice a year, you could make and sell goat's milk, soap or lotion made with the milk, or delicious artisan cheeses for local restaurants or markets.
- Meat—The market for goat meat is growing across the continent. In 2007, Canada imported over 3.5 million pounds of goat meat. The United States imported over 22 million pounds. Estimating an average carcass weight of 32 pounds a piece, we can approximate that 800,000 goat carcasses were brought onto North American soil for the year.
- Fiber—Goats produce mohair and cashmere used in high-end clothing and accessories. The majority of market production of mohair in the United States originate from Texan farms. Creating a sizable income from the fiber market alone requires a substantially sized herd and specialized knowledge. The secondary meat market of less productive animals is strong.
- Pets—Goats can be great pets for both adults and children. The standard-sized breeds are common in the pet market, but the popularity of pet miniature goats is increasing. These pint-sized personalities can easily be kept with little more than a large backyard to explore. Should you decide to breed and sell miniature kids, you'll find prices comparable to those of a purebred puppy.

ABOVE: Don't overlook the flashy Nigerian Dwarf goat as being just a pretty face. They are excellent milkers, have the most multiple births, and produce a high-quality milk that is perfect for making delicious cheese.

Main Goal: Dairy

Raise dairy goats with the added benefit of meat once per year.

By purchasing two bred dairy does, you will have your first taste of farm-raised milk as soon as the kids are born. If you stagger their breeding each year, say four months apart from each other, you'll keep the milk flowing for your family.

Doelings could be kept to increase herd size or sold as potential milkers for other farms. Bucklings could be sold to others to raise for meat, butchered before weaning, or grown on for another six months at minimal cost.

Main Goal: Meat

One of the most popular strategies involves breeding dairy does annually to keep them in milk. The kids are then raised for just a few months when they can either be sold or butchered and consumed.

The second most popular is to purchase weaned kids in early summer to raise for meat. By the time winter rolls around, your freezer is full of low-cost but highly nutritious meat. Purchase two young wethers for this strategy. If available in your area, get Boer or Boer-cross wethers for the highest yield.

When you're raising kids for meat, your goal is to raise the largest possible kid, in the shortest period of time, at the lowest possible cost. The least expensive goat meat to raise is from a six- to eight-week-old kid. Since dairy kids will only weigh about thirty-five pounds at that age, you will only net fifteen to twenty pounds of freezer meat. If time, money, and space permits, you could grow him on for a larger yield. By twelve weeks of age, he will weigh fifty pounds and will have only cost you a few dollars in grain and hay. Put him on pasture (with normal supplemental feeding) and by the time he's reached seven to eight months, he should weigh in at eighty pounds; about thirty-five pounds for the freezer. These are all estimations based on a 45 percent dressing percentage of goats.

The exceptions to the rule of age-weight increase are the Boer and Angora breeds. Boers are larger at birth and grow faster than any other goat in the same amount of time. The twelve-week-old kid example from above could easily be twice the weight of a dairy or grade kid at the same age. Angoras grow at slower rates than the meat or dairy breeds and do not reach their full weight until their second birthday.

Main Goal: Fiber

If you'd like to raise goats for fiber and have no interest in raising kids for meat or milking does daily, two or more wethers are ideal. Wethers are very friendly as a rule and yield a higher fiber count than does.

You could also purchase two bred Angora or cashmere-producing does to have a little milk, the fiber you desire, and offspring that eventually feeds your family.

Of Dairy Breed Interest . . .

- Nubians produce milk with high butterfat content and are often referenced as the Jersey equivalent of cattle. They are also considered to be the most vocal.
- Saanens and Sables produce the highest volume of milk. They are compared to Holstein cattle. Their milk is lower in butterfat than the other breeds, which makes it more palatable to unconverted fans of cow milk.
- The Toggenburg goat is the oldest registered breed of any animal in the world, with records tracing back to the seventeenth century.
- If you are more interested in milk than the meat from the offspring, consider a miniature version of your favorite dairy goat; almost all breeds are available as minis today!

Dairy Goat Breeds

Of the six common dairy goats, the Swiss breeds (Alpine, Oberhasli, Saanen, Toggenburg) are the hardiest for colder climates. The remaining two (LaMancha and Nubian) are genetically equipped to handle extremely warm and dry climates but may be kept in the north with proper care.

The majority of dairy goats in this country are managed much the same as dairy cows. They require milking twice daily and have a 305-day lactation cycle. This allows a 60-day dry period used by the doe to centralize her energy into the growing fetus and to replenish her body's store of nutrients. When a doe has her kid (freshens), it is common practice to remove the kids and raise them separately from their mothers.

Alpine

The Alpine is one of the larger dairy goats and a popular breed for commercial dairies. Discovered in Switzerland, the Alpine breed gained quick favor across Europe. Alpines match the Saanens in milk quantity.

This breed is known for her amiable personality. She has a straight or slightly dished nose and erect ears. Hair is medium to short. Most often seen in variations of black and

white or brown and white. Other color patterns include chamoisée (any shade or mixture of brown, often with a black stripe along the back and white markings on the face), two-tone chamoisée (usually a lighter color on the forequarters), cou Clair (a light-colored neck), cou blanc (a white neck with black rear quarters), cou noir (black front quarters and white hindquarters), sundgau (black with white facial stripes, white below the knees and hocks and white on either side of the tail), and pied (broken with white, spotted, or mottled).

Alpine does should be at least thirty inches tall at the withers and weigh at least 135 pounds. Bucks should be at least thirty-two inches tall and weigh at least 160 pounds.

Also referred to as the French, Swiss, British, or Rock Alpine.

LaMancha

The most striking feature of the LaMancha breed is ear formation. Ear shape is rated as either a gopher ear or an elf ear and should be no longer than two inches by breed standards. Gopher ears contain no cartilage, just a ring of skin around the auditory canal. Elf ears con-

tain a small amount of cartilage and a small amount of skin, which could turn either up or down from the cartilage. For buck registration, only gopher ears are allowed.

The origin of the LaMancha is a controversial subject amongst breeders. Some say this breed is a direct descendant of a breed imported to the United States from Mexico in settlers' times; others say they originated in Spain. No concrete documentation of their origin has been found to date. The LaMancha breed was officially recognized in early 1950s in the United States.

LaManchas have consistently pleasant temperaments and are considered by many to be the calmest of the purebreds. Hair is short, fine, and glossy and can be any pattern, color, or combination of colors. They are top producers of high butterfat content milk.

LaMancha does should be at least twenty-eight inches tall at the withers and should weigh at least 130 pounds. Bucks should be at least thirty inches tall at the withers and should weigh at least 155 pounds.

Nubian

The Nubian goat is the most popular dairy goat for commercial dairies. Considered to be the most vocal and energetic of the breeds. Nubians will thrive in almost every North American climate with some consideration for extreme temperatures and weather conditions. Their milk is high in butterfat (5 percent or more) and are an excellent choice for small farms interested in cheese making.

The Nubian breed originated in the United Kingdom and is popular with both British and Near-East cultures.

The most striking feature of the Nubian are their large pendulous ears, which are both long and wide. They have a strong convex profile. Hair is short, fine, and glossy. These moderately large goats are found in many shades and marking patterns of brown.

Nubian does should be at least thirty inches tall at the withers and weigh at least 135 pounds. Nubian bucks should be at least thirty-two inches tall and weigh at least 160 pounds. Also known as Anglo-Nubians.

Oberhasli

Oberhasli was once classified and registered with the Swiss Alpine and American Alpine breeds. Although not as common as some of the other dairy goats, this breed is growing in popularity across the United States due to their stunning appearance and calm, gentle temperaments.

Their heads are broad at the muzzle, tapering to a thin nose. Foreheads are wide with prominent eyes. Ears are short and erect and point forward. Face shape can be dished or straight. This breed is most commonly found in hues of reddish brown with accents of black along their dorsal, legs, belly, and face. On occasion, a fully black kid will be born from a breeding of two Oberhasli.

Oberhasli does should be at least twenty-eight inches tall at the withers and weigh at least 120 pounds. Bucks should be at least thirty inches tall and weigh at least 150 pounds.

Also known as Swiss Alpine, Oberhasli Brienzer.

Saanen and Sable

Saanens and Sables are registered separately but differ only in the color of their coats. The Saanen is white, the Sable taupe. This large dairy breed produces an average of four to six quarts of low butterfat milk per day.

Saanen and Sables have forward-facing, slightly erect ears. Faces and noses are narrow. Profiles can be dished or straight. Both sexes are bearded. Amiable, gentle, and calm natures.

Saanen and Sable does should be at least thirty inches tall at the withers and weigh at least 135 pounds. Bucks of these breeds should be at least thirty-two inches tall at the withers and weigh at least 160 pounds.

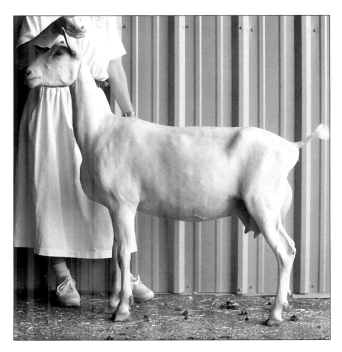

Toggenburg

The Toggenburg is the oldest registered breed. The Toggenburg breed is small in stature and produces a low butterfat milk (2–3percent). With pleasing personalities and a small frame, this goat is relatively easy to manage.

Coloration ranges from light fawn to a dark chocolate. Toggenburgs have white markings, white ears with a dark spot in middle, white stripes from above each eye and extending to the

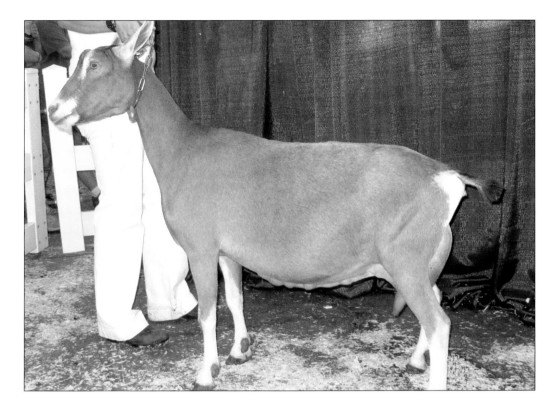

muzzle, white socks from hock to hoof, and a white triangle on both sides of the tail. Wattles and beards are common for both sexes. Profile can be straight or dished.

Toggenburg does should be at least twenty-six inches tall at the withers and weigh at least 120 pounds. Bucks should be at least twenty-eight inches tall and weigh at least 145 pounds.

Meat Goat Breeds

Quick Facts on Goat Meat

- Australia exports over sixteen thousand metric tons of goat meat annually. Over 50 percent of that export comes directly to North America.
- Goat meat is the most common red meat consumed worldwide. North American market demand is growing. Increased ethnic diversity represents the greatest demand for goat meat but health conscious consumers are quickly discovering the many uses for this lean and tender protein source.
- Cabrito, the meat of a ten- to twelve- pound milk-fed kid, is a considered a delicacy among the Hispanic culture. The United States Hispanic community is currently over thirty-five million strong. This community is forecasted to comprise 18 percent of our total population by 2025.

- African, Asian, Caribbean, Filipino, Jewish, Italian, and Middle Eastern cultures also purchase goat meat for standard dinner fare as well as religious and ethnic holidays.
- The Roman Easter market prefers goats that weigh 20 to 50 pounds. The Greek market, between 55 to 65 pounds. The Muslim market, 50 to 70 pounds and 100 to 115 pounds. The Christmas market has the widest range: 25 to 100 pounds.

Although any breed of goat (including the grade or scrub goat) can be used for meat, this classification is specific to the few breeds who grow the quickest and have more lean muscle volume than other classes.

The Spanish and Myotonic goats have been raised for centuries by North American farmers to provide a reliable meat source for their families. Both grow to decent proportions and are well muscled animals.

In less than twenty years (beginning in 1993), three new meat goat breeds have entered and dominated the North American market. The South African Boer breed, the New Zealand Kiko goat, and the Savannah breed. Although the Boer has been the most favored, the Kiko is gaining ground due to their excellence in high meat-to-bone ratio. Savannah goats are relatively new to the scene, but interest among southern goat farmers is strong.

Boer

Boers were developed in South Africa to grow quickly, produce flavorful meat, clear over-grown land, and thrive in inclement weather. This thick-boned and heavily muscled breed

BELOW: Purebred Boer Buck, CSB Ruger Reloaded Ennobled

is capable of grazing during all seasons and will grow larger in less time than any other breed.

Boers are commonly seen with white bodies and reddish brown heads, but solid varieties and spotted types are rising in popularity. The Boer is mild tempered and easily trained.

Boer does range from 200 to 250 pounds, bucks from 240 to 320 pounds.

Kiko

The Kiko is the product of a New Zealand government-funded initiative that began in the 1970s. The crossbreeding program involved both wild and domestic goats with the intention of producing a fast-growing meat source. Within a few generations and only minor variations during development, the Kiko was established. Kiko translates as "meat for consumption" in Maori (the native tongue of New Zealand).

Kikos are now proven to have the highest occurrence of kidding twins, a fast and reliable growth rate, and an unmatched aversion to disease.

A Kiko may not grow as heavy for the scale as the Boer but it does have a higher meat to bone ratio, therefore providing more consumable product per pound. Kikos are generally white but color variations can also enjoy full-blooded pedigree.

Kiko does average 100–150 pounds, bucks from 250–300 pounds.

BELOW: Purebred Kiko Buck, Penn Acres Huckleberry

Spanish

The Spanish goat has been present on American soil since the days of early exploration. These goats were either left behind or escaped and became feral. Through evolution they have grown climate tolerant and well muscled. For these reasons, the Spanish goat is often used in crossbreeding programs across the country.

Between current crossbreeding practices and numerous feral generations, the Spanish goat is found in all color variations, face shapes, and weights. The most notable characteristics are horn curvature and ear carriage. Ears are long and wide like a Nubian ear, but carried horizontally and slightly forward. The breed is currently a conservation priority by the American Livestock Breeds Conservancy.

Spanish goats are often incorrectly labeled as scrub, brush, or grade goats (terms used to define goats of no known heritage).

Spanish does grow to an average of 150 pounds, bucks to 250 pounds.

Savanna

The Savanna goat is the newest introduction from South Africa. This breed is hardy and suntolerant with a temperament similar to the Boer. Popularity and interest is growing for this new-to-the-scene breed.

ABOVE: Purebred Savanna Buck

Savanna goats are usually pure white with solid black hides. With a roman nose (rounded face) and long pendulous ears, they resemble a Nubian in the head shape, but a Boer in body type.

Savanna does grow to an average of 130 pounds, bucks to 240 pounds.

Myotonic Goat

The Myotonic goat has a genetic propensity to myotonia congenita. Myotonia is a muscular response to a startle. The response results in a spasm that locks the animal's legs, potentially causing unbalance similar to a human fainting. Brain and nervous system activity is unaffected by the myotonic response. This response condition also renders the animal helpless in the pasture until the animal regains control.

Myotonics were introduced into America by a Canadian (Nova Scotia) breeder. Repeated stiffening has given this breed muscular thighs—enough to be classified as a meat goat. Raised more as a novelty in modern times, the Myotonic breed was listed in the threatened breed category by the American Livestock Breeds Conservancy until just recently.

This is an intelligent, friendly, and easy-to-keep breed of goats with minimal desire to climb over or jump fences. Available in a wide variety of colors and coat lengths, these goats are straight-faced with long, slightly protruding ears and bulging eyes.

Myotonics are the smallest breed in the meat class (seventeen to twenty-five inches tall at the withers). Does should reach 130–150 pounds. Bucks have been noted to reach 200 pounds.

Also known as Wooden Leg, Stiff Leg, Nervous, or Tennessee Scare Goat.

Fiber Goats

Two distinct fibers are produced by goats—mohair and cashmere. While mohair is only produced by the Angora goat, cashmere can be found on over sixty of the world's goat breeds. (In North America, we most often find cashmere on Spanish and Myotonic breeds.)

Finding a top-producing cashmere goat is difficult in North America, not to mention an expensive acquisition for the small herder. Primarily you won't be looking for a breed, but a type (or class) of goat. For a goat to be classed as a cashmere, it should produce fiber with a crimp, be under 19 μ in diameter, and over 1 1/4 inches in length. The size of μ, from a simplistic understanding, is the result of a calculation from an equation of two or more measurements.

Angoras, on the other hand, are a pure and registered breed. These silky-haired goats generate ten to fifteen pounds of mohair annually. A wether produces slightly more than a doe.

The Angora is certainly in a class of its own, but the care is similar to that of other goat breeds. If you are only planning on keeping a few and hope to sell the fiber twice a year, find

a local breeder or Angora Goat group to collaborate with. Not only will they share invaluable breeding, feeding, and coat care tips, they will also assist you in finding a buyer for your yield.

Angora

An import from Turkey, the Angora breed has been in North America for over 150 years. Texas is the second largest producer or mohair worldwide. The breed is neither hardy nor an easy keep in comparison to other breeds, but they are an amiable pet for knowledgeable and conscientious owners.

Angoras are a silky-haired breed with long coats of wavy or curly locks, five to six inches in length. Although most Angora goats kept for fiber collection are white, they are also found in black, red, and brown.

As focus is on coat growth and quality, their diet should be closely monitored. Prone to parasitic infection given their dense coats and seldom set out to browse in mixed pasture where seeds and burrs can collect in the coat, lessening the value.

Does grow to an average of 95 pounds, bucks to 190 pounds.

Miniatures

Miniatures are currently popular on farms with limited space or need for full-production animals and their output. When I decided a few years ago to settle on a breed, I almost passed over any goat labeled as a mini, dwarf, or pygmy until a goat-keeping friend convinced me to consider the yield and personalities of these breeds.

Although only two miniature breeds are widely available in the United States and Canada, more than a few crossbreeds are worthy of a second look.

The Pygora (a cross between an Angora and a Pygmy) produces a lesser grade mohair but will also produce the fine down a Pygmy provides and finish larger if you are interested in meat. The Kinder (a cross between a Pygmy and a Nubian) is a great dual-purpose crossbred for both milk and meat.

Finally, almost every dairy breed has a miniature version now. If you run a search on the Internet you'll find Mini Manchas, Mini Toggs, Mini Nubians, and more. Many of these miniature versions provide a better match of production quantities for the average family!

The miniature breeds are perfect for family farms, make great pets, and are easy enough for children to handle. These goats are 1/3 to 2/3 the size of a full-size breed and require much less space and feed.

Nigerian Dwarf

Considered both a miniature and a dairy goat, the Nigerian Dwarf doe can provide up to a gallon of milk per day at her peak, but she averages a little over a quart by standard. Small- to medium-sized families interested in putting high-quality milk and cheese on the table

BELOW: Purebred Nigerian Dwarf, Hidden Hollow Wysteria

would be hard pressed to find a better match than the Nigerian Dwarf. Butterfat content in milk ranges from 5 percent on average to 10 percent near the end of lactation.

The Nigerian can also breed year-round and is known for having multiples—twins, triplets, and quadruplets.

The face of the Nigerian Dwarf can be straight or slightly dished. Their coats are straight and medium length and come in every pattern and color imaginable. Most are naturally horned and some will have blue eyes. They are friendly, easy to work with, and bond quickly with their keepers.

By breed standards, Nigerian Dwarf does should be at least seventeen inches tall at the withers, bucks at least nineteen inches, and neither should be taller than twenty-one inches. Mature does average thirty to fifty pounds, bucks and wethers forty to eighty pounds.

African Pygmy

The Pygmy Goat, another African import, has gained wide popularity as backyard pets due to petting zoo exposure. They are hardy and adaptable and make excellent little browsers for overgrown fields and small pastures.

BELOW: Purebred African Pygmy, Monterey Bay Equestrian Center

Although it isn't documented, there are two breeding styles for Pygmy owners. While some will breed for the purpose of pet stock, others are bred to be milked. You will need to have small hands to milk a Pygmy, but at her peak, a doe can produce two quarts per day of a high butterfat milk (approximately 6 percent). The milk is higher in calcium, iron, potassium, and phosphorus than the average dairy breeds' milk and lower in sodium.

The small stocky legs and lack of agility of the pet Pygmy makes containment less challenging than other breeds. Coats are straight and of medium length and come in a variety of patterns and colors.

By breed standards, does should be no taller than 22 3/8 inches at the withers, bucks no taller than 23 5/8 inches. Weight averages are fifty to eighty-five pounds.

I Just Want One

Please get two. You could have more than two, but you should have at least two. Every goat needs a companion.

Having witnessed multiple testimonies regarding single-goat ownership I agree that some goats are *capable* of living alone. The question, however, is not if they are capable, but if they will thrive in your care.

Nature dictates, by the very essence of the beast, that a goat should be part of a herd. Your goat has a long ancestry of wandering plains, mountains, and deserts with at least a few of her own kind. A horse, cow, or ewe will make a fair companion animal, but to pair them up this way is to stray from thousands of years of instinctual programming. You are not ever likely to see a goat partnered with a horse or cow, wandering together in the wild.

Animals that are designed by nature to live in a herd will experience stress when forced to live without their own kind. Internalized stress often manifests as illness, compromises immune systems, and lowers production. Stress will also affect longevity. Goats need goats.

Where to Find Your Goats

If a particular goat hasn't already caught your eye and captured your heart, it's time find the best stock that suits the goat shed you've created, the role you'd like them to fulfill, and your budget.

One of the best places to find local breeders with third party opinions is your feed supply store. The staff or owner of your local store know the people who purchase goat ration and medications. They should be genuinely happy to help you out; you are a potential new customer after all. Be sure to elicit their opinion on the breeder. You want to find the person in your area who has the best goats and is a knowledgeable and attentive owner.

Goat owners are fanatics. We love to talk about goats, our favorite breed, other people's goats, and to help out new goat owners. Good owners are happy to give advice on all aspects

of care at any hour of the day or night. Each one that you meet can open up an entirely new network of nearby breeders and goat husbandry associations.

While you are at the feed store getting a referral, check the bulletin board for advertisements. If you don't see a goats-for-sale sign, be proactive and write up a quick "Goats Wanted" ad and post it to the board before you leave. On your way out the door, look for a farm newspaper or magazine distributed by, or available at, your feed store.

Your goats could be just one click away! Over the last few years, the Internet has grown enough to easily find nearby breeders or associations for every breed. Type your state or province, plus the breed you seek into a Google browser window. You should find hundreds of listings.

Goat Speak: More New Terms You Need to Know

On the day you buy your goat, you might hear some new terms.

- Registered Purebred—Comes with a traceable pedigree (much like a purebred dog comes with registration papers and a family tree).
- Advanced Registry—Pertaining to milk does. This goat has been noted and registered as supplying a decent volume of milk over the course of a year. Dependent on her current age and health, Advanced Registry does have a proven record of milk production.
- Star Milker—Pertaining to milk does. The star system is based on a one-day test by volume.
- Grades—A grade goat may or may not be a purebred animal. It is without papers and registration. If your goat meets certain requirements and you desire it, you might be able to register it as a "recorded grade" with the issuing authority in your area.
- American, Experimental, Native on Appearance (NOA)— These are also grade goats. Americans and Experimentals are the result of crossbreeding programs and are not purebred animals, but a NOA might be.
- Polled—A goat without horns. The goat may have been born that way or had the horns removed later in life.

Bringing Your New Goat Home

Any change in a goat's surroundings and routine will cause stress. Know your goat's current feed program (right down to the very hour) and bring a week's supply of her previous ration and hay home with you. For a few days, don't alter her old routine, then slowly switch her over to your farm's hay, grain, and any supplements. This is true whenever you are making changes to a goat's feed or routine; make the change gradually over the course of a few weeks.

The seller should supply you with the following:

- registration papers (if applicable)
- veterinarian contact information
- list of past medications and vaccinations
- feed (grain, ration, and hay) for the first transitional week
- hooves trimmed and horn buds removed (if applicable)

ABOVE: Although this doeling looks shy peeking out behind the tree, she is just playing coy. As long as she'll let you catch and hold her at this age, you should have a good match for your farm.

Goats won't take up much room on your farm. Aside from their feed bins and a small platform to sleep on, their housing requirements will be minimal. In many cases, a large shed will work fine for a few goats. With ingenuity and a little room in your budget, you can have the ideal setup for keeping goats.

Housing Systems

Oh Those Poor Freezing Goats!

Dairy goats will be at their best when temperatures are between 55 and 70 degrees Fahrenheit. Milk production and feed consumption levels experience noticeable changes only in the extreme temperature ranges—above 80 degrees and below 0 degrees Fahrenheit. With only a few concessions in care, the goat is capable of coping in temperatures below freezing.

Dairy goats don't sweat. Dependent then on the climate in which you raise them, the challenge is often greater to keep them cool in the summer than it is to keep them warm in the winter.

There are two primary methods to housing and containing goats. The first is to pasture them and provide a poor weather and bedding shelter. This is method of choice for the hardy meat breeds or for large herds of dairy goats from spring until autumn. In regions where the weather dips well below freezing for days on end, goats will need a little more protection.

If the pasture area is too far away from their full-service shed or barn, you'll need to provide a pasture shelter. This doesn't need to be much more than a 4' to 6' high, three-sided enclosure with a slanted roof. Allow 8' to 10' square feet per goat in the pasture shelter.

The other method, loafing and confinement, is the act of keeping goats in a shed or small barn with a fenced yard for exercise.

The loafing and confinement system of raising goats is used mostly for dairy breeds or by farmers who don't have ample pasture. Sufficient room is provided for the goats, both inside and out, for sleeping, eating, fresh air, sunshine, and exercise.

The act of confinement keeps high energy activity to a minimum. Less energy expended by the animal allows for more productive use of feed. Angoras nearing a shearing date are often kept this way to control parasitic infestations and organic matter getting caught in their long hair.

The average dairy goat requires only twenty square feet of living space indoors, plus two hundred square feet of exercise space outdoors. Meat goats require more: thirty square feet inside, three hundred outside. Large herd meat producers only provide fifteen square feet inside as the animals prefer to be on pasture for all but the worst weather. Miniatures require a third less than dairy breeds.

I personally wouldn't confine a goat to such a small amount of space, but I have the luxury of acreage. I have had friends who didn't have much space but gave their goats a very nice life nonetheless, and I'm sure you could do the same if the situation called for it. Fresh air, cleanliness, clean water, and compassion make far happier lives than twice the space and an uncaring owner.

Shelter From The Storm

Goats are happy to spend their days and nights on pasture with the bare minimum of cover. Most breeds are comfortable between temperatures of 40 and 70 degrees Fahrenheit, but none of the breeds like to be wet. Meat breeds may continue to graze in a light summer rain but should have a shelter to return to at all times. Forced exposure to rain in low temperatures often results in illness. Kids and low-condition individuals are at the highest risk for respiratory infection and hypothermia.

Extreme Weather Temperatures

Not one of the domestic breeds common to North America are built to handle long bouts of below-freezing temperatures. These goats, unlike other livestock, do not grow enough of a

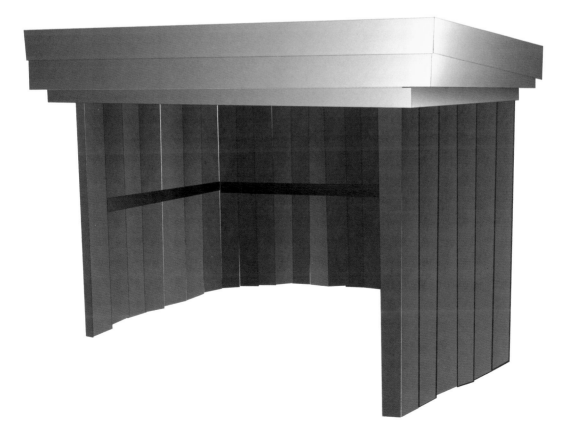

LEFT: An outdoor sheltered feeder keeps supplemental hay off the ground and protected from the elements.

thick coat or put on a layer of fat directly beneath their hide to prepare for the full thrust of winter. They will keep themselves warm by huddling together and through energy expenditure through rumen activity. Adequate shelter, herd mates, and sufficient hay make the coldest winter nights bearable.

Dairy goats that are used to being locked in for the night should be kept in draft- and moisture-free buildings with ample dry bedding. Take extra care during extremely cold temperatures if kidding is imminent or if you already have kids in the barn due to early-season births. Extra bedding, a supervised or safe heating unit, and a little extra hay to eat will help keep the harsh chill from bothering your herd.

Meat goats that are kept on pasture year round should have their shelter turned in such a way that the prevailing winds aren't blowing into the open side of the shelter. More bedding should be added (daily if necessary) to ensure that the goat's extremities aren't touching frozen ground. More hay will also be required to replace the energy spent on staying warm.

If at no other time, cold days and nights are the perfect time to treat your goats to warm drinking water. Goats actually prefer slightly warm water in all seasons so the investment of safely installing an immersion heater or a water bucket with the heater core built right in would make the cold season more liveable.

During dry or hot summer months you'll find goats equally resilient and capable of managing in extreme temperatures. They may move around less, spend more time in the cool shade of the barn, and drink more water to stay comfortable. If you lock up your goats at night, you'll want to make an exception for the hottest nights unless you have plenty of windows that provide a cross breeze.

With all the extra time spent in barns or shelters, litter and bedding need to be changed more often. Goat owners can't afford to miss the extra work required during these times. Urine and feces quickly build high ammonia levels, which can cause respiratory health complications. Kids are at the highest risk as they are closer to ground level where ammonia is strongest.

Converted Sheds

If you will just be raising a few goats and you already have a shed on your property, you might be able to convert it to a nice little barn for your goats. Conversions save you time and the cost of building a brand new structure. Knowing the number goats you will house at the height of the season (your does plus offspring that you keep for a few months) will determine if an existing building is adequate.

When I decided to move my dairy does out of the communal barn and into their own space, we took an old 10' x 12' shed and spruced it up instead of starting from scratch. Fresh wood went on the walls, and a small slab of concrete was poured onto the milking area floor. After a few partial walls, a manger, and a sleeping platform were added, the goats moved in.

In a 10' x 12' shed, the communal stall will take 35–50 percent of your floor space. This leaves adequate room for a milking station, feed storage, and one or more smaller stalls. You'll need at least one small stall for isolation of a sick goat, quarantining a new goat, kidding, or weaning.

Shed Conversion Example

Two dairy does given 40 square feet (20 square feet each) for a communal stall, plus a smaller stall for kidding (20 square feet), a section for milking and grain storage (35 square feet), plus 20 square feet extra for two kids annually.

At less than 120 square feet required, a little planning can convert a 10' x 12' shed into an adequate loafing barn. Tight, but adequate.

If you're building from scratch, design your floor layout to accommodate the feeding and watering of goats without entering their communal stall. The easiest way to do so is to build a half wall between their space and yours. Your side contains the manger, water bucket, and soda/salt feeder. Their side contains slatted or key-holed head access to all three. Open areas (above the half walls for example) can be securely sealed with 12" x 12" page wire fencing material.

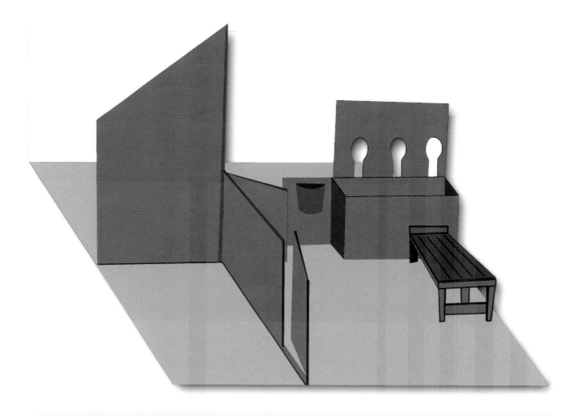

ABOVE: A converted 10x12' shed. Showing a milking stand, keyhole manger access, exterior water buckets, and good distribution of space: 60 square foot communal stall, 20 square foot kidding stall, and 35 square feet of work and storage area. Add an access door to the back wall that opens to their yard and you're all set to bring your goats to their new home!

Indoor Mangers

Designing and building a slatted or key-holed access manger may be extra work right now, but it will save you money in the long term. Goats are not only notoriously picky about the hay they eat, they are also the most wasteful. If the manger is open, you will soon discover that goats take a mouthful of food, turn to see who might be behind them every few minutes, and drop half of their mouthful on the floor. Once it has hit the floor, they will have very little interest in it.

Goats also love to climb into the top of open mangers. They will stand there and eat, curl up and fall asleep, trample the hay with dirty hooves and perform body elimination functions. Once soiled, no self-respecting goat will touch the hay.

The standard top width of a keyhole is eight to nine inches with a keyhole shaped taper to the bottom at four to five inches wide. The height of the keyhole from top to bottom is sixteen inches. When the opening height is optimal, goats will crane their necks to put their head in at the top and then lower their heads to a comfortable fit within the slot.

ABOVE: Two different styles of head-only access for feeders. The one on the left is traditionally made of plywood; on the right, rough 2 x 4 lumber.

If the top of your wall is higher than most of your goats can reach through slats or keyholes, you could build a variable height platform on their side of the wall.

Keyhole entries won't work for horned goats. They won't be able to fit their heads in, and if they somehow manage to do so, they could get stuck and panic when they are finished eating.

A slat-access manger built high on the wall will work for both polled or horned goats. Width between the slats should be about six inches—just enough for a goat's nose to push through and grab a mouthful of hay. If you can build it with a hinged rear opening, you will be able to fill it from your side without fighting off hungry goats.

Floors and Bedding

Dirt flooring in a goat barn with a thick layer of bedding and litter material to absorb smell and moisture is adequate. Straw and waste hay are easy and economical, but wood shavings are easier for cleanups and more beneficial as future compost.

Goats will choose an area of their barn for sleeping unless you provide an elevated platform against a wall. They won't naturally keep this area clean, but it will stay cleaner than the rest of their living space if placed in the corner or in a low traffic area.

All sleeping areas should have thick bedding that is changed or added to regularly. Lay fresh bedding over the existing every few days and completely replace bedding material when soiled, damp, or smelling like ammonia. If at any time you can smell ammonia in the goat barn, make deep litter removal your top priority.

Waste Removal and Garden Compost

The rich organic waste material removed from the barn makes great garden compost. Goat droppings and litter will not burn your plants or delicate root systems if you were to use it right away on flower beds, but be sure to compost it for a six month minimum before using on vegetable gardens. If you've been using waste hay as bedding material or litter, I suggest composting it under a dark tarp or bin for at least a year or you'll have a miniature hay field sprouting in your gardens from all the seeds.

Lighting Increases Production

My first dairy goat shed didn't have lighting, and as the days grew shorter, I found myself milking and feeding in the dark more often than not. One simple light made all the difference to my comfort level, and it also kept the does in top production. Combined natural and artificial lighting for eighteen to twenty hours per day maintains milk production through the fall and winter months. It also increases the chance of an early spring estrous.

Running Water Saves Many Steps

If it is a special treat to have a water tap in the barn; it is ten times so to have both hot and cold water there.

If you are keeping dairy goats, you will save countless steps carrying milking paraphernalia back and forth to the house twice per day. Those trips become even more annoying when you or your doe have knocked over or dirtied one of your cups or buckets and you have to return to the house, resterilize everything, and start over.

Hot and cold running water in the barn isn't a treat once you've had it for a few months, it's a necessity. You can properly wash up in the barn. You can start the cooldown process immediately for raw milk. You can give your goats the slightly warm water they love to have in cooler temperatures.

Safe Grain Storage

The idea of a goat breaking into the grain or feed bin, suffering from bloat, and dying before I have a chance to help is one of my biggest fears. It happens more often than you'd think, which is why I now store our livestock's grain in the basement of the house. It is convenient—the basement door opens to the back of the house, which is where I begin my chore journeys each morning and evening anyway.

Most people store their grain directly in the barn so it is on hand for on-the-spot ration adjustments or refills if one of the containers get knocked over.

Grain should be kept away from all moisture, out of the sun, off the ground if it is bagged, and definitely out of a goat's reach. A galvanized trash can with a snap-on lid placed well out of a goat's reach is perfect. They are cheap to purchase, last for decades, and also keep vermin out of your grain.

Feeding Styles Based on Goat Type

As feed will account for the majority of ongoing costs associated with keeping goats, allowing them to pasture or browse will lighten the financial burden considerably as long as it suits your purpose.

Just as there is a best breed for your intended use, so will there be a best method for keeping and feeding that breed. A dairy doe, for example, would not be set out into overgrown fields of raspberry brambles and spiked saplings as she could damage her

BELOW: Goats are a relatively easy keep. You can set them on pasture, in overgrown fields, and even partial forests. As long as there's enough to eat, they'll thrive.

Brush and bushes like this thorny weed are soon banished from your fields. Although the goats may not eat the entire plant in one pass, eventually, the weeds that decrease the value of your fields will just stop growing.

udder. Furthermore, some wild plants will unpleasantly alter the taste of her milk. You would be better served to keep that doe on a field of managed pasture full of nutritious grasses.

As with every rule to keeping farm animals, there is an exception to the pasture-for-dairy-goats rule. If you plan to make cheese year-round, a confinement system of shed and yard might be the better option as does fed mainly on lush green pasture will produce a thin, lower butterfat milk.

Goats raised for meat or fiber are the most versatile on the types of terrain they can find sustenance on. The one exception is the Angora breed. Mohair from the Angora goat will net higher returns when it is free of twigs, leaves, and bugs picked up from time spent in fields or forest.

There is just no way around it: the opportunity for an introduction of illness, parasites, toxic reactions, and bacterial infections are everywhere on a farm. At times the cause will be within your preventative power, at other times the cause will be out of your control.

Being overly protective of your goats is not a character flaw. While those without goats may scoff at the idea, compassionate keepers will feel justified in their sensitivity, and the logically minded will recognize this as simply taking care of what is yours. No matter how you justify attentive care, we are their keeper and their health is our responsibility.

Here are my top 10 tips to keeping healthy goats.

1. Only buy from responsible sellers.
2. Only buy healthy animals.
3. Quarantine new goats for at least thirty days.
4. Strive towards cleanliness in the barn. You can't (and shouldn't) sanitize daily or pick up every piece of waste matter as it falls, but ensuring that fresh water is always available and bedding is dry and clean are two easy tasks that simply can't be neglected with goats.
5. Create and adhere to a vaccination schedule on your veterinarian's advice.
6. Be regimented in your schedule. Goats are creatures of habit and lovers of schedules—especially the dairy breeds. Lactating does know when the milking hour draws near, and they'll often be waiting at the barn door twenty minutes before your regular time. If you provide daily rations of grain, you'll see this same behavior in all breeds.
7. Block out times in your personal planner for routine maintenance such as hoof trimming, annual lice and tick checks, and body condition assessments.
8. Prevent illness related fatalities before they start. If you can spot a warning sign (such as low energy or lack of interest in food), you can take appropriate action before her health deteriorates further.
9. Know the common ailments of goats.
10. Whenever possible, keep a closed herd. Goat owners who are regularly bringing new goats in, participating in shows, and visiting other farms have a higher risk of stress-related and airborne illness in their own barns.

The goat is a seasonal breeder.

As the days grow shorter, the doe begins to look for a suitable mate. Scientifically, we say she is seasonally polyestrous, but in layman's terms, we simply call her a short-day breeder. As a rule, she won't be very picky about breeding either. This makes her owner's task of finding a suitable mate easy to accommodate.

Aside from choosing a buck for her, there will be equally important decisions to be made—namely when she's bred, the best time to have kids, and what to do with all those extra mouths to feed.

Animal husbandry is never an exact science; the act of procreation even less predictable. As I write this chapter, I can think of a gross exception for every time span I've quoted, every breed I've mentioned, and every situation I've warned against. What follows is to be considered a guideline that loosely covers the many breeds and classes of North American goats.

Unless you plan on only growing wethers that you purchase from a local farm, you need to be well armed with knowledge, and in control of the breeding process. From buck selection criteria to due dates, being in control of your doe's breeding is as important as her supply of daily fresh water.

ABOVE Although goats are only slightly uncomfortable in temperatures just below freezing, it will be easier on your farm budget if kids reach their wean date after the snow has left.

The typical breeding season runs from September to February. At our farm in central Ontario, I hold off breeding does until late October to early November with an expectation of kids in late March and early April.

Late and staggered breeding dates result in late and staggered due dates across my small groupings. There have been times when a few kids are born in the same evening, but I usually have time to carefully watch over each doe while she kids. Where spring takes its time arriving, breeding late is precautionary. Kids are not born into a subzero Fahrenheit barn months before the pasture is ready for them. Late breeding also saves me money. By the time kids are ready to eat solid food, it will be growing just outside their front door.

When to Breed

Even though your doe might think she is ready to breed at six months of age, don't allow it until she is eight to nine months old. A better rule of thumb is to wait until she has reached at least 75–80 percent of the average mature size for her breed. Early breeding could result in a stunted doe that has trouble kidding. In subsequent years, she will also have less chance of multiple births.

The other option—letting her miss her first year of breeding—fares less well according to some breeders. Waiting a full extra year will not only cost you money without return, but it will also create a doe that is a poor milker and an inattentive mother. They advise that even if she hasn't hit the right weight on the scale, breeding young is the better option.

Dairy goats are bred every year. In the warmer regions of North America, does are bred during September and October. Then after a five-month-long gestation, they give birth right before spring, when fresh pastures will soon provide the most nutrition.

Meat goats are usually bred as often as possible, every eight to nine months, with the optimum breeding time between August and October.

Angora goats are bred, like dairy does, once a year. This is usually between August and November, right after a fall shearing. By arranging and controlling this goat's breeding, we are assured that they'll be ready to kid shortly after the spring shearing.

Goats raised for cashmere are bred anytime before November. This practice helps all owners to wean kids by mid-June and dry the does off. Lactation, it seems, slows the annual growth of new cashmere.

ABOVE Choose your buck based on genetics and past performance. This spotted Boer from Rebel Ridge Farms in Virginia was fashioned through an intelligent, managed breeding program.

Is Your Doe in Heat?

Your doe may not exhibit all of these behaviors, and she may even have a few of her own.

- Vocalization. Some will bleat loudly as if in pain.
- Wagging the tail more than usual.
- The under tail area may darken, appear swollen, or be wet with mucus.
- Dairy does will have a change in milk volume. High volume right before the heat and less volume during the actual heat.
- Could show a decrease in appetite.
- May show an increase in urination.
- Will pace fence lines, sometimes to the point of exhaustion.
- May mount other does or let other does mount her.
- May have a slightly swollen and reddened vulva.

A hoof trim is one maintenance task that doesn't take much time, but is an absolute necessity. Without proper hoof care, your goat can become sick, lame, or permanently crippled. Frequent attention to the matter, no matter how minor, is less time consuming and far easier than infrequent major trimming.

From the age of two months, I make all goats stand and allow inspection of, or a slight trim of, their hooves every month—no exceptions. Doing so at such a tender age helps them get used to the feel of the shears and file and establishes who is the boss in our relationship.

If you pasture dairy goats on somewhat rocky pasture or in a field that has centuries-old rock fences, the hoof will stay trim a little longer. Goats raised on a small yard could be given a small concrete slab or a stack of large rocks to climb on to help wear the hoof wall down. These aids may minimize the amount to be trimmed but will not excuse the monthly trim.

Our meat herd spent a substantial amount of time on rocks and rough terrain, and there were always a few with overgrown hooves by the month's end. Hooves, like hair, grow at different rates on different goats—they all need to be checked regularly.

To Trim a Hoof

Confine each goat in a manner that prevents escape. A dairy stand with stanchions, a leash and collar tied to a barn wall or fence, or having a helpful friend with you are all great ways to keep a goat still for the amount of time it takes to trim hooves.

If your goat is being particularly difficult, gently lean your shoulder into her side with her other side against a wall while you work. Angoras and kids can be "sat" on their rumps, leaning their backs against your legs while you work.

Stay calm while you work, and you will find that your goat remains calm as well.

1. Soft hooves are easier to trim than dry hooves. An hour in a yard with morning dew will expedite the task.
2. Bring shears, a rasp, and hydrogen peroxide to the barn with you.
3. Wear gloves.
4. Constrain the goat in a manner that allows you to easily and comfortably bend her leg for observation of, and work on, the entire bottom of the hoof.

The Anatomy of a Hoof

Step 1: Gently scrape out any impacted dirt in the curve of the hoof with a hoof pick or the point of closed shears. You are working gently inside the wall. Impacted dirt at the toe of the hoof does not need to be forced out. Get the debris that falls out easily as you trim. It will all have fallen away by the time the hoof has been fully trimmed.

Step 2: Snip away the overgrowth at the toe of the hoof (it will be the longest), then move onto the sides and remove any folds or excess found there. Your objective is a level foot—a hoof wall that just barely extends and protects the frog. The center area of the hoof, the inside wall, can also be trimmed and should be marginally smaller than the outer wall. This ensures that the weight of the goat is on the outer edges of the hoof.

Step 3: Use a hoof rasp to create a smooth finish on the hoof wall. You can also shave or plane the bottom of the foot if it is overgrown with a hoof knife until you begin to see a pinkish tone on the base of the inner hoof.

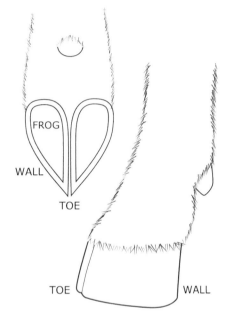

ABOVE: A nicely trimmed foot. The base of the hoof is parallel to the Coronary Band (the area where the hair ends and the cuticle begins).

③
Pigs

WELL-CARED-FOR PIGS ARE quick to trust and bond with a keeper. They are unmatched by any other farm animal in this regard. They will look at you with soft and round eyes, sing while they eat the meal you've prepared, and grunt responses when you talk to them. Within minutes of passing over the barn's threshold they'll have you by the heart-strings, and in no time at all you've fallen in love with that snouty little face. Therein lie both the joy and the hardship of raising a pig for consumption.

Contrary to popular belief, a pig's preference is not to be dirty or to eat anything that lies in his path. Instead, he is a barnyard animal tidier in his habits and more particular about his food supply than the vast majority of modern pedigreed dogs.

Pigs have been raised by small and large farms for centuries. They are intelligent and social animals completely capable of showing appreciation and learning simple commands. In most parts of North America, small farmers like you and I will raise a spring pig or two for a year's worth of pork, bacon, and roasts.

The pig arrives at six weeks of age when the air is just starting to warm and departs shortly after the first leaf falls off the maple tree. This is an age-old tradition built on climate, the fast growth of the pig, and temperatures cool enough to hang and chill pork.

With minimal space and bother, a pig or two could find happiness on your farm if only for a few short months. They are easier to raise than chickens if you are aptly prepared and understand their nature before you bring one home. Without this knowledge you run the risk of being outwitted by a barnyard animal, but you won't be alone. They have outwitted the best of us. Even the experienced farmer can be caught by surprise from time to time.

The Benefits of Farm-Raised Pork

Raising pork for your family's meat supply is rewarding, economical, and gainful. The finished pork excels in both flavor and texture and the versatility of the meat will satisfy even the most discerning diner in your family. Fresh, cured, or smoked; made into sausages,

> **Meat Yields for Pigs**
>
> Very little of a farm-raised pig is wasted: 75 to 80 percent of live weight will have value.
>
> One six-week-old, 40-pound piglet plus five months of quality care equals one 220-pound pig.
>
> One 220-pound pig butchered equals 140 to 175 pounds of fresh pork, smoked bacon, and ham roasts plus 20 pounds of lard.

chops, ribs, or roasts; fried, barbecued, or braised—the possibilities of pork are limited only by your recipe books or imagination.

You'll also have an unfair advantage over the grocery store shopper. No longer will the pork on your plate be seasoned, cut, and packaged based on market demand, but by your specifications. Would you like extra pepper or garlic in your five-to-a-pack sausages? Mesquite smoked hams instead of honey glazed roasts? Half-inch-thick chops or fast fry cuts?

Far more important than flavor are the health-related benefits in avoiding commercially raised pork. News reports of mad cow disease and foot-and-mouth virus, recalls on processed meats, and similar outbreaks have left many of us wondering about the safety of any grocery store meat. Uncertainty over hormone and antibiotic residues, inhumane care of the live animal, unsanitary living conditions, and sick slaughtering practices all contribute to less than palatable meals placed on your table. No longer will those issues be of concern.

When you grow your own pork, the decisions of medicating and feeding an animal for consumption are yours to make. You can rest assured that the animal raised on your farm was given the best care and led a stress-free life. That the meat your family consumes hasn't been fed a diet of garbage, such as restaurant leftovers scraped off plates and serving dishes—some of which might be spoiled, tainted, or rancid by the time it reaches a pig's trough.

All in all, raising a pig is a satisfying and economical way to fill your health-conscious family's freezer with delicious and versatile meat. The project takes a mere five months. Daily care requires only twenty minutes, lesser still if you opt to pasture your pig.

The Hardship of Raising Pigs

I would be remiss if I didn't aptly warn you on this one important aspect of raising a pig. At least 80 percent of people I have spoken to, who have raised a pig at one point in their lives, have struggled over this very issue because they were not prepared.

A pig's time on your farm is short. Before you even begin to set up a pen or fence a pasture, certainly before you start shopping around for piglets, you need to ascertain that you and all family members will keep your eyes on the road ahead or face an emotional struggle (in some cases deeply scarring) as the day draws near for the pig to leave the farm.

A sow, having reached over five hundred pounds in her first year, is fit only for breeding and costs a small fortune to feed every month.

Although it is noble to grow in fondness for the animals you keep, keeping a pig past the date you've set for his leave is neither economical nor feasible. A pig grown to maturity can easily top the scales at seven hundred pounds. He will require more food daily than a family of four and the housing or containment of such a large animal can be a nightmare. At seven hundred pounds, when you've decided you just can't keep him anymore, he will not be of optimum use to a butcher and his life will have been in vain.

For children involved in raising pigs, many of life's lessons are explored. A child under the age of six has the most potential for having her heart broken and may vow never to eat meat again. Certainly not pork.

Should emotion overpower logic and sending your first farm-raised pig off to the butcher is particularly difficult, console yourself in the certainty that you provided a better life for the pig than a commercial grower would have on all accounts.

Deciding on a Breed

Eight main breeds of pigs are raised in North America (Yorkshire, Duroc, Hampshire, Landrace, Berkshire, Spotted, Chester White, and Poland China), but unless you are planning on raising a breeding pair, the registered and heritage breeds are of little importance. All are similar in their efficiency at feed conversion and growth rate before one year of age. Most of the piglets available for seasonal growing are purposely crossbred for feed conversion, quick growth, and resistance to stress. It has been said that these pigs grow 10 to 15 percent faster and use feed 5 to 10 percent more efficiently than a purebred.

Worry less about breed when choosing your pig and more about temperament, signs of health, overall weight, and body construction. I look for long, lean torsos in piglets and usually end up with a cross of a Yorkshire or Spot. In the past ten years many varieties and crosses have entered my farm, and each one I've raised has been as satisfying and prolific as the last.

Choosing a Good Pig

Healthy piglets have a smooth, sparse coat of hair. The skin underneath is bright and clear with no flaking or protrusions. Eyes are bright and active and noses are soft and moist or dusted with the dirt they've just been rooting in.

There is no escaping the signs of a healthy piglet when you meet one. They are alert to each other, the people at the fence gate, and their general surroundings. Any new action is investigated by curious piglets without apprehension. Young pigs who are disinterested or inactive are to be avoided at all costs.

Spring pigs, also known as weaner pigs or feeder pigs, should weigh in between thirty-five and forty pounds at six weeks of age. Any piglet weighing less than thirty pounds should be overlooked—even if you have success raising this small pig he will be less economically viable.

Although good breeding practice can determine the potential of a piglet, if it hasn't had a good start it will never reach that potential. Contrary to popular children's stories of excellence in runts, a piglet abandoned by or removed from his mother before turning six weeks

A lone piglet appears to be staring at her reflection hoping for some company on the other side of the water. Healthy piglets are intensely interested in the world around them.

of age will be adversely affected in growth and health for the remainder of his life. The task of hand-raising a runt piglet is a serious undertaking not worthy of the average person's time or investment.

The castrated male pig will grow a little faster and finish ten to twenty pounds heavier than his sister. Other than this minor difference the two will be indiscernible. If you have preselected a male pig, ensure he has been castrated and the wound healed before you bring him home. Intact male piglets are not a bargain at a livestock auction. You will either need to castrate the piglet yourself or have a tough time raising him (boars can be aggressive from a very young age) as well as ending up with pork that isn't fit to eat.

Designing Your Small Farm Strategy

People who have raised pigs for years will tell you that they grow faster, larger, healthier, and happier if they have their own kind about them. I've not seen valid research on this theory, but knowing the nature of the pig, I buy into the concept every year.

If your family doesn't have a need for two butchered pigs in the freezer, buddy up with a friend or two and split the costs and chores. Even in our very small circle of friends, there are always one or two who jump at the opportunity to have farm-raised pork in their freezers come fall.

Although a pig's feed is inexpensive when compared to the amount of meat he provides, you can save between 30 and 40 percent of your feed bill by growing him on pasture. This strategy also removes the chore of waste removal and will assist you in returning any overgrown area of your farm to a pasture state. Having raised both sorts, I now prefer the pasture-raised pig. They have consistently been the happiest, friendliest pigs with the best-tasting meat.

A Pig's Temperament

A pig is an inquisitive animal with a sharp memory. Quick to figure out the operation of latches, levers, or valves, they will use their talents to create a surrounding that suits their needs. This could be as innocent as puddle creation but as devious as an escape.

A pig will also attempt to amuse himself if bored. Having another piglet to grow with or toys to play with will keep him out of trouble and mentally stimulated. He'll happily push around large sturdy balls for hours in the pasture, as well as enjoying a tire swing to butt with his snout.

ABOVE: Pigs can be raised alone as long as they have ample company from humans or another farm animal, but in my own experience they do far better with at least one of their own to grow with.

If the transition to their new home has been easy, piglets make
quick work of rooting up their yard for a cool spot to lie in.

Although a pig can be raised individually, their very nature dictates otherwise. Pigs are social animals. It just never sits right with me to see one being raised alone in a pen at the back of the property, given no toys, company, or attention except for required twice daily visits for fresh feed and water. Yes, raising just one pig can be done successfully, but at what quality of life to the sociable pig?

Mentally Tethering a Pig to Home

When you first bring a piglet home, spend a few weeks getting to know each other in the barn or in a nearby outside pen. This gives him the opportunity to bond with family members and associate this new, safe place as home. Doing so mentally tethers the pig and is invaluable to the pastured pig farmer as well as owner of pen-raised pigs. This will be the place to which he'll return should he ever break free.

Should your pig escape, he isn't likely to go too far once he has bonded to you or to the barnyard. He will, however, get into someplace you won't want him. Flower gardens are favorites as are children's wading pools.

Call a pig back into the pen with a tempting snack and fix the fence while he eats. If he is stubborn about going back in, you can move him with a few good smacks on the butt. If he is adamant about staying put, consider staying with him to ensure he doesn't get into too much trouble. When he's sufficiently hungry or decides that he's discovered enough for one day, he'll return to the place you've prepared.

Some farmers never have any trouble with pastured pigs, pens, or fences. Could it be that they just know how to pick a good pig and build a great enclosure? Or might it be that all of a pig's needs are met within those confines and he has no desire to escape? The answer lies more within the second question. Pigs—fed and watered adequately, given a shelter that is full of soft bedding and overhead protection, provided with ample space, company, or toys to alleviate boredom, a wallowing hole to cool off in, and gentle summer breezes blowing through the area—will rarely test the fence. Would you?

Pig Pens and Pasturing

A six-week-old, fully weaned piglet will miss the warmth and security of his mother and littermates. He is optimally kept in an area that maintains a temperature of 85 to 90 degrees Fahrenheit for at least three weeks. A heat lamp or heater directed toward his bedding area is all that is required. At sixty pounds, your piglet will have grown enough to manage cooler evening temperatures of 65 to 70 degrees. Extra bedding material and a second piglet can make up for a drop of another 5 to 10 degrees.

Bedding and Temperature

Take extra precautions to keep your pig's bedding area dry and draft-free. The only exception is to allow air circulation in the shelter for older pigs being raised in a southern climate.

Bedding should be four to six inches thick. A pig likes to burrow in when he sleeps. Straw or wood shavings are acceptable materials.

Wet living conditions will make a pig sick. Always place a pig's shelter on high ground.

Outdoor Housing

Established pigs, as rough and tough as they appear, always require protection from the elements and a roof over their heads when they bed down for the night. The roof keeps heat in, and rain and sun out. Light-skinned pigs will burn without protection in summer sun. If possible, keep them near a stand of trees so they will have shade without having to be in their shelter. A wallowing hole is of great use to the light-skinned pig in hot summer months.

Shelters are constructed with three-foot-high walls on at least three sides and a rain-proof roof. The shelter must be large enough for the pigs to completely stretch out in. Most pigs prefer to sleep together.

Make the shelter sturdy enough to handle the pig's weight—he will be flopping around inside as well as rubbing up against the outside to scratch himself.

Bedding should be changed when soiled and kept at a minimum height of four inches when replaced. Pigs like to burrow in for the night—warm, dry, and partially covered by their bedding material.

If you build their house with a base it will be easier to move to pasture or rotational pasture.

Wallowing Zone

A pig's joy over mud holes and puddles is more about staying cool than a passion for muck. If you have the room, allot an area in their pens or pasture for a large puddle.

ABOVE: Pigs enjoy puddles more than they do mud, but either is welcomed as a means to cool themselves off on a hot day.

If you don't provide a wet zone they will attempt to make their own by spilling drinking water every time you turn your back. These highly intelligent and inquisitive animals will also figure out, in short order, how to operate auto-mated watering devices. They'll play with valves and levers until they've created a puddle large enough to wallow in.

The Penned Pig

A pen must be, at the very least, spacious enough for a pig to change direction without bumping into the perimeter.

Well aware that a world to explore exists outside the space you have allotted for them, pigs will look for a way out. If their living conditions are inadequate or even if they are just lonely, they will even risk pain and personal safety. Once your pig has learned to knock down an enclo-sure, door, fence, or gate, you can count on a summer of trouble and "chase."

ABOVE: Heavy-duty cattle panels may work fine to contain larger pigs that have lost their interest in discovery and escape.

You'll need to fully understand a pig's ability, strengths, nature, and habits in order to contain him.

Pigs love to lean, shove their noses in, and rub their bodies on every object. Any barrier to freedom must be sturdy enough to sustain over two hundred pounds of repeated pig pres-sure—pushing, thumping upon with front hooves, bumping into, and rocking.

If they can't go through it, they might go under it. Soft ground at the base of a fence or gate is an invitation to root and wallow. After rooting into soft ground and pushing their noses through to the other side, they are smart enough to know that freedom and adventure are just a few inches away. Digging under the fence is less than a day's effort—worthwhile work for a pig bent on escape.

Pigs who don't feel they are getting their fair share of room, are bored, are not having their basic needs met, and have learned that fences were made to be broken will consistently test the fence.

Piglets, especially those who have not had sufficient time to bond to their keepers, will have you running to catch them more than a few times until you learn to outwit them with a pig-proof enclosure. A tightly woven wire fence works well when supported by sturdy posts, as will a four- to five-foot-high solid wood enclosure. Keep an eye on the base of any outdoor enclosure—a piglet can make short work of digging under a fence to get to freedom.

Some farmers use barbed wire in conjunction with wood fencing to contain their pigs—nasty stuff that can cut and create scar tissue on a young pig. Barbed wire is no longer economical in comparison to the new electric fencing units. Solar-powered units can be purchased for less than the price of ten spans of barbed wire.

Finding Piglets to Purchase

In the spring, keep your eyes peeled for classified advertisements in your local newspaper. You might also find a flyer at the feed store. If not, ask the store clerk for a few names of people who raise and breed pigs in the area.

When you visit a farm with the intention of purchasing a pig, you are checking for overall care and cleanliness of the barn and pens. You should hope to find a seller knowledgeable about pigs and happy to supply information such as birth date and weaned date, and have dewormed the pig the day before you take him home. He will also supply you with a few days' worth of feed so you can slowly switch your new pig over to your own preferred brand. Happily pay an extra 10 to 20 percent for a piglet whose origin you know, rather than one you may find at auction.

Many piglets are also purchased at the livestock auctions. Often these piglets are sold as full litters from one sow or a mix of runts from different sows. You'll find each in groups of five to eight. Every so often just a pair will come into the ring at a time, which is perfect for those of us looking to raise one or two for our family meat supply.

Some nice litters have been known to appear at auction houses. If finding piglets to purchase has been a challenge and you are fortunate to find a litter of healthy piglets at auction, you'd be well advised to purchase the entire lot— even if you only want one or two! You should have no trouble reselling healthy piglets. It is often the case that even before you load too many piglets into your pickup truck, you'll be approached for a sale of two or more. In that same manner, don't be shy. Ask the person who outbid you if they'd like to sell one or two of the lot. They may say no, but if they say yes your search for piglets is over.

In smaller country auctions you might be able to find out who brought the pigs in for sale. If you can, ask the seller the same questions you would have asked at the pig farm as well as the brand of feed the pig has been on. Answers to each will prove invaluable once you get the piglets home.

Should you find yourself back home from auction with more pigs than you need, write up a quick flyer or newspaper ad. You'll be down to the number you originally wanted in no time, even turning a profit in the process. Be sure to ask for more than what you paid. You had to invest the time and costs of travel, and perhaps duel in a bidding war to obtain them.

BELOW: Piglets selling at a livestock auction are often sold by litter. This group appears to be a good size, attentive to their surroundings, and clear of illness.

What to Feed a Pig

Not much different than you and me, a pig is healthier and grows to have more muscle than fat if he is given lots of exercise and fresh greens. This is but one of the reasons a pastured pig is healthier and tastier than a penned pig that is fed nothing but carbohydrate-laden grains all his life.

Water

Pigs overheat in temperatures rising above 80 degrees Fahrenheit. Drinking fresh, cool water keeps them from dehydration as well as heat exhaustion. During summer months, even while he's young and small, a pig can drink two to four gallons of water throughout the day.

Although there are many types of mechanical and automatic water devices available, if you're on a budget you can use any watertight, food-grade, plastic barrel sawed in half or a durable, wall-mounted bucket available from the feed store.

Whichever setup you choose, you'll quickly find that a pig loves to spill, walk through, and root under it or operate valves and nipples in order to make a wallowing hole in which to cool off or play. If your system is neither automated nor secured, be prepared to make a few trips to the water daily to refill the supply.

Water should be changed daily and the bucket or trough cleaned whenever sullied. Once a week it doesn't hurt to give a mild bleach rinse to all feed dishes, buckets, and troughs.

Feed

ABOVE: Pigs on pasture should receive at least one meal of pig ration to round out their diets and should have fresh water accessible at all times.

Even pigs at pasture require a daily meal of pellets or mash to ensure their diets are complete. Penned pigs are happiest when their meals are split into two portions through the day. Commercial pig feed is the most economical and nutritionally balanced food for pigs at any age.

Recent reports on the health-related issues of commercially raised meat are based on medications and chemicals found in feed—the same feed you and I serve our home-raised pigs. Many people make the mistake of substituting corn for commercial feed as it is a natural product. Corn, however, will do little for the pig other than to put on fat. If the problems associated with medicated feeds are a concern, check the label at the feed store and ask for alternatives. It may take a few phone calls to find a supplier of nonmedicated feed.

Young pigs will gain one to one and a half pounds per day on commercial feed, at average ratio of 2.5:1 feed-to-muscle conversion. The average home-raised pig decreases his feed-to-meat conversion efficiency after 220 pounds.

The younger the pig, the more protein he'll require for growth and weight gain. Start your pig on a commercial starter mash that contains all the vitamins, minerals, and protein required. Then, regularly check his weight with a hog tape or by using the formula in the sidebar and switch to new feeds as set out in the table below.

Start young piglets with free-choice feeding, checked three or four times per day, until ten to twelve weeks of age, when you can switch to two feedings per day. You want to provide enough feed to ensure your pig is getting ample nourishment, but never so much that it is wasted. Spillage or spoilage, even a little a day, makes raising pigs far less economical. The table below will assist you in determining the average amount of feed required per day.

Feed Quantities and Protein Requirements for Pigs

	Pig Weight	Protein Required	Feed/Day	Consumption
Starter Grain	40 to 75 pounds	16–18 percent	3 pounds	60 pounds
Grower Grain	75 to 125 pounds	13–14 percent	5.5 pounds	140 pounds
Finisher Grain	125 to 220 pounds	12 percent	6.8 pounds	340 pounds

Weighing Pigs

Weigh your pigs using this formula: HG x HG x L / 400 = weight

HG (heart girth): Measure the pig all the way around, just behind the front legs.

L (length): Measure from the center, between the ears, to the base of the tail.

Multiply HG by HG by L and divide the total by 400. This is your approximate live weight.

If the approximate weight you've arrived at is less than 150 pounds, add 7 pounds. If the approximate weight is over 400 pounds, subtract 10 pounds for every 25 pounds the pig weighs.

Pigs are very susceptible to stress brought on by changing location, being transported, and changes in feed. Limit the effects of stress by transitioning them slowly to new feed. Over the course of a few days, gradually mix new feed in an increasing percentage to their previous feed.

Many farmers supplement their pigs' diet with table scraps and garden waste. While this may be economical, as with pasture-raised pigs, you still need to supply a commercial ration to ensure their nutritional needs are met. I never feed a pig meat of any sort, but I have read that many people do. Raw meat is the worst as it can carry disease that not only could affect your pigs' health but also your own after butchering.

An Old Farmer's Tale

When I first started raising pigs at home I was a single woman new to living in the country but not to horror stories. I was living on seventeen acres down a back-country road with no friends in the area, which ultimately meant that if something should happen to me, no one would know until the bill collectors showed up!

An old farmer told me when I bought my first pigs, "Never feed 'em meat. They'll turn mean."

The thought of having two 250-pound mean pigs to manage was enough to deter me from ever allowing any scrap of meat to enter the trough!

Keeping Pigs Healthy

As long as your pigs arrived healthy on your farm and have been vaccinated, dewormed, and kept with care you should have no reason for concern over their health.

Keep in mind that a pig loves to eat, so if one suddenly goes off his food you can safely assume that a veterinarian call is warranted. Other signs to watch for are covered in the previous discussion of the signs of an unhealthy piglet (see Choosing a Good Pig on page 76).

Internal Parasites

Every farm animal is prone to internal parasites and worms. Pigs are no exception. Piglets are usually wormed one day to a week after weaning and before being sold. If your piglet is six to eight weeks old, it will be your responsibility to worm him twice more at thirty-day intervals.

As with all medications, take the time to read the directions and follow them to perfection. Medication residues remain in the meat tissue of a pig for quite some time. Every brand of medication differs; therefore information regarding dispersal of residue must be printed on the label of all worming medication for all animals raised for consumption.

Butchering and Preparation

You can't say I didn't warn you. For the last four to five months, you have diligently cared for this inquisitive, intelligent, and trusting animal. He likes his back scratched. He hums to you while he eats the meals you empty into his trough. In many ways he has become a pet, yes, but always a pet with a purpose.

If you've been feeding your pig finishing grain from 120 pounds on (the last six weeks or so) and he has reached a reasonable weight of at least 200 pounds, it is time to say goodbye. The finishing grain and pasture ensure that the meat will be tender, but you can further sweeten it up by feeding your pig as many autumn apples as he cares to eat. You might like to cut them into bite-sized chunks to ensure the core doesn't somehow become lodged in his throat.

If he's been on pasture, bring him in and pen him up for the last three days on your farm. For the last twenty-four hours, don't feed him, but be sure to give him ample fresh water to drink.

If you can't find it within yourself to slaughter or you are not physically strong enough to manage the task, call around to local butchers to inquire about bringing in your live pig. Some slaughterhouses also offer pickup services. For a nominal extra fee they will truck your live pig off the land. In a few weeks they will return with endless packages of butcher-wrapped meat ready for your freezer (having chilled, smoked, and cured all cuts).

I hasten to say this is not the optimum situation, as your pig will suffer some stress from being transported to a new location where the smell of death hangs in the air. However, with trusted people performing the task and your specific instruction, you can rest assured that the matter will be dealt with as acceptably as possible. You can tell a lot by the way a handler loads your pigs onto their trailer. Seldom is rough-handling necessary.

Doing It Yourself

In keeping with the farmer's creed, "If you grew it to eat it, you'd better be man enough to kill it," I'll walk you through the process of slaughtering a pig. This is tough work that requires physical muscle, so don't attempt it alone unless you know you can lift or hoist 240 pounds of dead weight over your head.

You can cure, cut, smoke, and grind the meat yourself, but that is far beyond the scope of this book. What you'll find in the next few paragraphs is the involved and humane practice of killing your pig on his own land, and preparing two sides of carcass for a butcher to further process.

First and foremost, be sure to make arrangements with the butcher for the next day. You cannot leave this point of contact for the last minute as he or she may have important information regarding government regulations that you must abide by if you employ his services. You also want to ensure that he is expecting you the morning after slaughter. The last thing you want is to show up with two sides of fresh pork in the back of your truck only to find the butcher cannot accommodate you.

The evening temperature is also important, as you'll need to hang the carcass overnight in a cool (35 to 45 degrees Fahrenheit), clean, bug-free location while you wait for the butcher shop to open. Pork does not require aging, but it must be thoroughly chilled before cutting. Your butcher will hang and finish chilling the carcass. Curing and smoking will require another week or longer.

When you go to the barn, have every item you'll need at hand. A rifle, a sharp eight- to ten-inch knife, two meat hooks (a rope or chain will do in a pinch), a strong rope or chain (to attach to the meat hooks), running water, a pulley to act as a hoist (if available and required), a few large buckets, and rubber gloves. Two people will also need three hours uninterrupted.

I wouldn't suggest taking this on the first time without the help of a skilled and registered hunter. Shoot to stun the pig at point-blank range in the center of the forehead. Don't waste any time, as you may have only wounded and stunned, not killed, him. He will fall to the ground. Roll him over on his back, stretch his neck as far back as possible and make an incision through the skin and into the throat, just above the breastbone. Cut downward and in, until the blade is under the breastbone. Finally draw your knife through and straight down to sever the main artery. The pig will begin bleeding out.

Make two incisions, one on each foot, between the tendon and the bone just above the hock and hang the carcass until it has finished bleeding out. Wash off the carcass with your hose and proceed to remove the skin in strips, three inches wide at a time. This will consume most of your time on the job.

Once the pig has been skinned, cut across the back of the neck, directly behind the ears. Cut through the gullet and windpipe, then pull down on the ears, pausing to cut from the ears to the eyes and then to the point of the jawbone. This keeps jowls intact but allows you to remove the head.

Place your largest bucket under the carcass now. From the original incision, pry apart the breastbone as you cut. Be mindful not to cut deep into the abdomen or intestines. Cut around the sexual organs to within half an inch of the anus. Cut fully around the anus and tie off with a small piece of rope. The whole mess comes back into and through the abdominal cavity and out the hole you've just cut. You may need to cut a few muscles and the gullet to release it into your bucket below.

Hose out the inside of the carcass. Remove and save any large pieces of fat including the flaked leaf fat to render for lard. Finally, cut through the backbone with a knife or saw and leave the two pieces to hang and chill.

Birds

WHETHER YOU WANT TO become an expert at bird identification or just get to know your neighbors a little better, your first step should be to learn a few basics about where birds live and the reasons behind their choices. Identifying a bird begins with understanding the ecosystem around you. Why is this bird in this place? Is it here to eat or to build a nest? Or is it just passing through? What is it saying? Answering these questions reveals a world like no other—an alternative universe of high-drama courtships, joyous births, bittersweet departures, and heroic battles for life—all taking place in the treetops and shrubbery of your own backyard.

Territories

Within a bird's range is its territory, the area that he has decided to call "his" for the season. Bird territories are usually temporary. Even for birds that return every season, boundaries shift and change as other species move in and out. Males spend a lot of time and energy staking their claim, defending it from intruders, and announcing to others the boundaries of their empire.

Birds often have to adapt to the course of human events: A forest taken down over the winter for a shopping mall may require a bird to look at a local park that has traditionally belonged to someone else. A marsh drained for a subdivision may require a sudden change in plans.

All your own small decisions about your yard or woodlot affect your visitors. We had a big oak in a central location that had died, and we began to grow concerned with its stability and its proximity to our house. Every storm caused a shower of brittle branches. We sadly took it down, understanding the damage we had done to our population of cavity-dwellers. On the other hand, the addition of cherry trees and blackberry bushes to our yard, although not particularly beneficial to our own food stores, became a boon to birds that had overlooked us before.

Layers of Habitat

Have you ever thought about why chickadees and warblers are in the same yard? Or whether those cardinals mind having those jays around? The fact is, while your yard seems like only one environment to you, to the bird there are actually several habitats represented. Every species has specific needs that can be met without interfering in others' food sources or nesting requirements. Nature has created an elegant system that allows you to host a multitude of different types of birds in your own small backyard.

The reason that so many birds can comfortably make a temporary home in your backyard is because every environment has a series of separate and distinct habitats that perfectly suit individual species.

The canopy

Sometimes called the *overstory*, the canopy is the highest vegetative layer in the forest. The canopy is filled with leaves from the forest's large, most mature trees. Extending 60 to 100 feet high, this habitat is rich with beetles, caterpillars, and leafhoppers that make their homes on the highest layer of the treetops, providing a rich environment for foraging. Birds like blue jays, owls, hawks, and eagles live and work in the top layer of the forest environment. Some songbirds like the yellow-throated warbler build nests here. The thick leafy canopy provides protection for the open nests of these birds.

The understory

The understory of the forest is a shady world of younger trees and shade-tolerant varieties. Only 50 percent of the total sunlight can get through the canopy to the understory. Just below this top layer of sun-gathering leaves, this habitat hosts a completely different environment for bird

LEFT: Your yard represents four distinct layers of habitat. Different bird species eat and live in treetops, the shady understory, in shrubs, or on the ground.

BELOW: Eagles and other birds of prey find treetop perches ideally suited for spotting fish and small mammals below.

The petite ruby-crowned kinglet gleans small insects from leaves and small branches.

species. The understory features the trunks of the mature trees that leaf out the canopy, as well as the smaller, less-mature trees of the same species. The rough bark of these trees has split apart and developed cracks. These trunks are home to a staggering variety of worms, beetles, ants, and other insects. That's what makes it the ideal habitat for woodpeckers and flickers.

Smaller cavity-dwelling birds such as nuthatches like this environment too. But instead of hammering away at the trunk itself, you can often see them scooting down the trunk headfirst, gathering the insect eggs, beetles, spiders, and small caterpillars that the wood-boring birds missed. Chickadees find a multitude of hiding places for seeds and insects in the crevices of the mature bark.

Varieties of shade-loving trees also make their home here. In my backyard in Virginia, this layer features varieties such as shadbush, sourwood, dogwood, and redbud, which attract a range of seed-loving birds.

The shrub layer: The shrub layer is categorized by leafy plants that extend no more than 6 or 7 feet high. Azaleas, rhododendron, and mountain laurels are common varieties in this layer. This habitat

ABOVE: A white-breasted nuthatch working down the crevices of a tree trunk in search of insect eggs and seeds.

ABOVE: Shade-loving trees like dogwoods and redbud help form the understory layer.

is the ideal feeding ground for many small flying insects. Gnats and blackflies are plentiful here in the spring and provide a tasty, high-protein diet for a variety of birds that are preparing for nesting. The rose-breasted grosbeak lives in the shrub layer, preferring to build its nest in shrubs 5 or more feet off the ground. Robins, chickadees, warblers, quail,

BELOW: The northern flicker is as comfortable with dead wood as its woodpecker cousins, but actually forages more often for food on the ground. You can often find it hammering on the forest floor in search of ants and beetles.

BELOW: This female northern cardinal is at home in the safety of this tangle of berry bushes.

ABOVE: Dark-eyed juncos are ground foragers, but watch for them perching on low branches or shrubs.

ABOVE: Tall grasses and leafy undergrowth provide a natural hiding place for this white-throated sparrow.

sparrows, finches, and cardinals also find wild and cultivated fruits, berries, earthworms, and insects such as beetle grubs, caterpillars, and grasshoppers in this layer.

The ground layer: This layer features plant varieties that bloom mostly in the spring, along with lichens and mosses. Dead logs also host bark beetles, larvae, carpenter ants, and earthworms, as well as an array of spiders, centipedes, and slugs. Ground-foraging birds find an abundance of food in this layer.

Other Habitats

Meadows and Grasslands

Some birds, such as the red-winged blackbird and the eastern meadowlark, along with bobwhites and field sparrows, enjoy the open-air market of the meadow. Grasshoppers, insects, and seeds make up their diet. This is also a great place for predators, however, so ground nesters like the bobwhite have to be constantly vigilant. They take the precaution of sleeping in groups and camouflaging their nests with grasses and other nearby vegetation. A great hunting ground for small birds and rodents, owl and hawk sightings are also common here.

Marshes, Wetlands, and Waterways

Some birds live and work only where there is an abundance of water. Marshes and wetlands are important habitats for these species, and a great place to bird-watch. Marshes generally support large populations of birds that are uniquely suited to waterside living. Because water

ABOVE: A flock of red-winged blackbirds congregate in an open pasture.

height varies from year to year, some nesting birds create clever floating nests that allow them to adapt to changes. You can expect to see herons and egrets, which have ample sources of food in the marsh environment. Look for their nests high above the water. Saltwater environments are homes to yet another set.

What Birds Eat

ABOVE: A northern bobwhite quail blends easily into a background of leaves and dry grasses.

Birds live and work where the food is, and they have to work hard to find all the food they need. Their high metabolisms demand fuel, and lots of it. Birds eat anywhere between 5 and 300 percent of their body weight every day. Tiny birds like chickadees and wrens consume at least half their body weight in tiny insects every day. Hummingbirds will eat up to three times their body weight in nectar every day.

Multiple species can live in habitats together because different aspects of the habitat provide for different needs. Likewise, different seasons provide different food sources, and birds have adapted to take advantage of those sources. Your yard is appealing specifically because of what is going on there in any given season.

ABOVE: One of the most common of water birds, the great blue heron makes its home wherever the fishing is good!

ABOVE: Eastern towhees are perfectly at home in thick undergrowth, surrounded by berries and small insects.

Spring

Most species are on the move in the spring when insect populations are beginning to explode. The abundance of available animal protein makes it possible for mating birds to prepare for nesting season. Many birds are specialists, relying primarily on a particular category of insect. Others are more opportunistic, and will accept a varied diet. Some of the favorite food options include:

Caterpillars: Many birds rely on at least some insect protein, and spring provides it in spades with caterpillars. Nuthatches, warblers, cuckoos, mockingbirds, and blue jays are among the species that help us by munching on tent caterpillars, but cardinals, grosbeaks, tanagers, and many others also love a meal of caterpillars.

ABOVE: Dragonflies are great insect predators in their own right, but are also a juicy prey for birds.

Mosquitoes and other flying insects: Some birds specialize in catching insects on the fly. Martins have long had a reputation for keeping mosquito populations under control, but several other varieties of birds also do their part. Swallows, flycatchers, phoebes, warblers, and waxwings are among the many birds that like mosquitoes.

Beetles, grubs, and spiders: Ground foragers specialize in these bugs. Blackbirds, bluebirds, thrushes, wrens, starlings, and towhees are among those that prefer this category.

Earthworms: Many ground foragers hunt earthworms, with the American robin and other thrushes among the best-known customers.

Water insects and larvae: Certain birds specialize in waterborne insects. Some birds hunt the air over the water for these hatches, while others will go into the water for their food. Kingfishers, crows, robins, flycatchers, ducks, and shorebirds are among the many birds taking advantage of bodies of water for their food source.

Bark insects: Woodpeckers, chickadees, nuthatches, wrens, and sap-suckers spend much of their time examining and picking at the crevices of tree trunks and branches for the insects that live there.

Summer

Many birds that have spent the spring eating insects and other animal matter in preparation for nesting are now looking to the fruits and seeds of summer. Although many songbirds will continue to collect insects to feed their growing youngsters, they may begin to seek out berries and other fruits for themselves. Others, such as the cedar waxwing, prefer a diet that is almost entirely fruit. In summer, watch for birds looking for the following:

RIGHT: Downy woodpeckers are small and acrobatic enough to take advantage of larvae clinging to thin branches and weed stems.

ABOVE: The American robin spends most of the spring and summer in search of succulent earthworms, supplementing its diet with fruit and berries.

BELOW: If mayfly larvae are lucky enough to hatch from the water, they must dry their wings before flying off, making them an easy catch for birds.

An abundance of summer fruit is sure to attract a variety of songbirds.

Beetles: Sparrows, bluebirds, orioles, and downy woodpeckers are just some of the birds that are happy to munch on beetles. Cardinals, grackles, starlings, and robins are all helpful in controlling Japanese beetles.

Cicadas: When summer is filled with the sound of cicadas, robins, grackles, and buntings will take full advantage of this plentiful food source.

Flying insects: Specialists in flying insects are now enjoying the peak of their food supply.

Shrub fruit: Blackberries, raspberries, blueberries, grapes, and strawberries will attract mockingbirds, orioles, and cedar waxwings, among many others.

Tree fruit: Cherries, Juneberries, and mulberries are also attractive to fruit eaters.

Autumn

Autumn brings its own set of treats just in time for birds that are preparing to transition back to cold weather. The harvest is on, and birds are ready to take advantage of the changing season. Hardy fall bushes and shrubs are offering up native fruits and nuts with higher fat content.

Berry-bearing autumn shrubs: Winterberry, spicebush, sumac, and Virginia creeper attract robins, bobwhites, kingbirds, catbirds, and fly-catchers.

Decorative trees: Dogwoods and Bradford pears produce plentiful fruit and make a great meal for bluebirds and grosbeaks, among others.

Wildflowers: Weed seeds are abundant now, and sparrows and finches make the most of them.

Late-blooming flowers and fall berries are a boon to migrating birds.

Late-blooming flowers: Salvia, hollyhocks, and lobelias are greatly appreciated by hummingbirds as they start their fall migration.

Ground insects: Fallen leaves create great ground cover for hiding a wealth of insects, grubs, and other tasty treats for your ground foragers.

Winter

Depending on where you live, winter brings a variety of challenges. For northern areas, snow covers much of the ground-foraging possibilities, and leftover fruits, berries, and tree seeds have been largely picked over. As you head south, where the weather is cold but snow cover is minimal, birds will still find opportunities in the conifers and tulip trees.

Now is the time when the ground cover and the shrub and tree assortment in your backyard may distinguish it in the eyes of the birds that are planning to stay for the winter. Conifer trees, standing clumps of ornamental grasses, and evergreen shrubs like boxwood and holly are inviting places for birds

ABOVE: The tufted titmouse is a master hoarder, usually shelling a seed before hiding it away for later use.

this time of year. In addition to providing shelter from the wind and snow, these areas also make great hiding places for birds looking to escape predators that are also on the hunt for food.

Leftover seeds: Trees, garden flowers, and weeds are thoroughly inspected for these leftovers.

Insects: Even in the dead of winter, insects are hiding in tree crevices, beneath dead leaves or loose bark.

Winter berries and fruits: Holly berries, dogwood, rose hips, and bittersweet vines that retain their berries throughout the winter are a boon to birds this time of year.

Conifer seeds: Chickadees, pine siskins, and nuthatches are among the birds that may seek the small seeds in pinecones.

Where Birds Nest

Nesting season is the most important time of the year in the bird world. Males search out mates, and, depending on the species, males and females may search out the best location for the nest together. What is considered a perfect building site varies dramatically between species. A number of birds, from large woodpeckers to miniature nuthatches, use holes in trees. Others want the nice crook of a small tree for their nest, while still others look for a penthouse location, as high in the tree as they can get. Some birds will reuse their nest for another brood, while others refuse to use the nest more than once. Still others are perfectly happy to "borrow" someone else's nest.

Nests are feats of remarkable engineering, and nesting materials include a wide array of materials. Twigs, grasses, and roots or vines are common building blocks. Some birds, like the petite hummingbird, weave intricate nests using thistledown, spider silk, and moss. Others, like the robin, create a multilayered nest of twigs and grasses with a binding coat of mud and a lining of soft, dry grass.

Although there are more varieties of nest-building techniques than you can imagine, most North American nests can be categorized roughly into these types:

Ground Nests

The simplest of nests, this is a shallow indentation scraped in the ground, often used by shorebirds or pheasant, grouse, and partridge. Most birds will scrape or stomp down an area with their feet, although terns will find a sandy location and rock their bodies back and forth to create the shallow depression. Some birds will line the area with grasses or dead leaves, while a number of ducks will actually "feather their nest" with soft down plucked from their own bodies. Because ground nests can be quite exposed to predators, nature put in a few safeguards. Both the eggs and the nesting female tend to take on a camouflage color that helps them blend into the surroundings. Babies are born fairly mature, allowing them to leave the nest quickly, as early as twenty-four hours after hatching.

ABOVE: The eastern meadowlark's ground nest is woven right into the landscape and camouflaged from view with an arched opening.

Cavity Nests

Cavity-dwellers include both birds that are capable of excavating their own holes in trees, and those who rely on existing holes to make their homes. Woodpeckers, which dig their own cavities, often line the bottom of their holes with wood chips before laying their eggs. Other cavity-dwellers, such as eastern bluebirds, which rely on finding a vacant hole, may line their holes with wood chips, leaves, moss, or animal hair. Cavity nests have the advantage of being warm, dry, and sheltering. Unfortunately, they are fairly easy for predators to access, so cavity-

ABOVE: A nest filled with baby swallows, awaiting their next meal.

dwellers have some ingenious deterrents. Cavity-dwelling birds may smear pitch or resin around the entrance to deter visitors.

ABOVE: This wood stork's nest may be used by the family for several years.

Platform Nests

The most recognizable version of this nest belongs to one of many large birds that raise their young high in the trees. These complex nests, also known as *aeries*, are built by eagles and ospreys of branches and twigs. Often placed in the highest tree in the area, these nests are built to last, as bird families use the same nest year after year. Other birds use the platform style too. Common loons and some waterfowl build platform nests on the ground, and still other waterfowl actually create floating platform nests.

Cup Nests

When most people think of a bird's nest, this is the style they think of. You can find cup nests all around your yard—perched in the crook of branches, tucked into one of your hanging planters, perched in a gutter, or in the eaves of your garage—any sheltered place is a possible building site. Built by most songbirds, these structures are made in a series of layers. The female starts with the outer walls. She forms this of grasses, leaves, and

RIGHT: A perfect clutch of American robin's eggs; a female may produce as many as three clutches per year.

ABOVE: A buff-bellied hummingbird sits in her nest of bark and grass fibers, suspended delicately from a branch.

twigs, often glued together with mud. Then she lines the inner layer with soft materials such as grass, feathers, or moss.

Different species have variations on this basic structure. The female hummingbird's delicate nest favors spider silk as her "glue," and is a miniature work of art. Crows, which follow the same basic blueprint, create large, messy versions. Swallows build cuplike nests out of mud, often attaching them to the sides of houses or inside chimneys.

Pendant Nests

One of the least-common nests here in North America is the pendant, tightly woven of grass and other flexible plant materials. It is a sack-like structure and hangs precariously off a small branch. Favored by Baltimore orioles, it is surprisingly safe from predators. It has a small opening at the top or side to allow access by the parent, but it doesn't allow easy entry by a typical nest marauder. On the other hand, hanging from this little branch has definite disadvantages, as strong breezes give this little nest a wild ride.

How to Identify a Nest

If you want to know who lives in a nest, it is probably going to be a matter of watch and wait! First, take note of where the nest is. This will give you an idea of what you are looking for. If the nest is in the cavity of a tree or in a nesting box, you have narrowed down the possibilities to typical cavity-dwellers. If it's in a tree, note the height, size, and shape of the nest. You may find eggs which can help you identify what type of bird is nesting there. But the best way to know for sure is to see the parents coming and going.

If you have noticed a nest in your yard and are curious about what lives there, be respectful of the inhabitants. Nesting is a nervous time for the birds, so don't do anything that will disrupt the process. Don't peek into active nests first thing in the morning or after dusk; that's when the mother is most likely to be home, and you don't want to scare her off. Carolina wrens regularly use my hanging planters for their nests. I am always amazed by how quickly these nests seem to appear, so I always take a peek before I start watering.

Identifying the Bird

Tyrant Flycatchers (*Tyrannidae*)

As the name implies, this family likes flying insects, so many of its characteristics are related to its ability to catch them. Its bill is pointed and has a small hook to help it snatch bugs from the air. Its demeanor is still and watchful—it has to be still in order to hone in on its next target.

- **Size:** Small to medium in size
- **Shape:** Large head with a high, rounded crown, strong substantial bill, short tail
- **Attitude:** Upright, sits very still, perhaps only moving head; holds tail straight down
- **Behavior:** Prefers insects; watch them dart out for flies and then return back to the same perch
- **Common family members:** Flycatchers, phoebes, eastern kingbird, and wood peewees

ABOVE: Flycatcher silhouette

LEFT: The eastern phoebe, a member of the flycatcher family, can be readily identified by his *fee-bee, fee-bee* call.

Kinglets (*Regulidae*)

His name means "small king," and he indeed reigns in the forest as one of its smallest members. This family eats insects and insect eggs, so its needs to be agile and acrobatic to forage off leaves and small branches. Watch for it to feed this way, that way, even upside down.

- **Size:** Very small, the smallest of the sparrows
- **Shape:** Short body with small, needle-like bill and an incised tail
- **Attitude:** Can be quickly identified by the constant flicking of its wings
- **Behavior:** A very nervous little bird, remarkably acrobatic and in constant motion
- **Common family members:** ruby-crowned and golden-crowned kinglets

ABOVE:
Kinglet silhouette

ABOVE: The regal golden-crowned kinglet is barely larger than a hummingbird.

One of the most identifiable members of the thrush family, the American robin has an air of confident authority.

Thrushes (*Turdidae*)

This rather large family is readily identified with its most well-known member, the American robin. Thrushes are generally ground foragers and have a pretty varied diet that includes earthworms and other insects, but they will also dine on fruit and berries. Take note that even though the bluebird is a member of this family, it has a slightly different posture and feeding style.

- **Size:** Medium in size
- **Shape:** Round-bodied with a rounded head, strong chest, and a strong multipurpose bill
- **Attitude:** Confident and alert, with an upright posture
- **Behavior:** Runs or hops along the ground. If ground foraging is interrupted, it generally flushes to a tree branch where it remains quietly until it can go back to what it was doing.

ABOVE: Thrush silhouette

- **Common family members:** American robins, bluebirds, and thrushes

Sparrows (*Emberizidae*)

Sparrows' plumage is generally muted in color. Shades of brown, gray, black, and rust dominate. They can easily be mistaken in the field for a finch or female birds of other species, so learn their shape and habits to help your identification efforts.

- **Size:** Small to medium
- **Shape:** Plump, round birds with short, rounded tails and small powerful beaks for crushing seeds
- **Attitude:** May cling to a plant stalk while it eats, even on the ground; sits upright in tree branches
- **Behavior:** Ground foragers, they can be seen scratching or hopping along the ground, looking for seeds or insects
- **Common family members:** Sparrows, towhees, and juncos

ABOVE: Sparrow silhouette

ABOVE: This white-striped variation of the white-throated sparrow is a common visitor at backyard feeders.

Finches (*Fringillidae*)

Bright colors and brilliant songs—the finch family is definitely worth attracting to the backyard! Mostly vegetarian, they seek out the tiny thistle (nyjer) seed that other birds often ignore.

- **Size:** Small to medium-large
- **Shape:** Large round head, compact body, strong conical-shaped bill, and a notched tail
- **Attitude:** A generally gregarious family—look for them to feed in congenial flocks. Largely vegetarian, they seek out thistle seeds but will eat other seeds
- **Behavior:** A bouncy flight pattern with a combination of flapping and gliding; they often call while in flight
- **Common family members:** Finches, grosbeaks, redpolls, pine siskins

ABOVE: Finch silhouette

ABOVE: Finches thrive on a diet of tiny thistle seeds.

Wrens (*Troglodytidae*)

Seldom still, you might first notice a little wren because it seems to be everywhere at once, checking everything out. Very comfortable with suburban conditions, it will often make its nest in a planter near your house rather than out in the woods.

- **Size:** Small
- **Shape:** Compact bird with an erect tail that is easy to identify; a flat head shape; and a long, thin bill with a slight curve that lets it pick insects out of holes and crevices
- **Attitude:** Jaunty and upright, with that distinctive tail
- **Behavior:** An energetic demeanor that is more "busy" than "nervous"; often seen hopping quickly around low shrubs and thickets
- **Common family members:** Wrens

ABOVE: Wren silhouette

BELOW: The Carolina wren's loud *teakettle, teakettle* song is easy to recognize.

Wood Warblers (*Parulidae*)

Sometimes referred to as "the butterflies of the bird world," both because of their tendency to flit nervously and because of their bright plumage. The biggest treat of springtime for those living in the East and Midwest, these small, colorful birds migrate en masse, sometimes arriving together by the hundreds.

- **Size:** Generally smaller than sparrows
- **Shape:** Petite and round, with a thin needle-pointed bill that can easily snatch insects. It's all about color with the warbler—most have distinctive markings, at least some in yellow
- **Attitude:** Too busy to perch for long, can be found hanging upside down, checking out the underside of a leaf
- **Behavior:** Nervous energy—they tend to dart constantly from branch to branch
- **Common family members:** Warblers, chats, yellowthroats

ABOVE: Warbler silhouette

ABOVE: A thrilling sight for fall birders is the migration of hordes of yellow-rumped warblers making their way to their winter homes.

Vireos (*Vireonidae*)

Easily mistaken for a warbler, these birds generally appear in gray, olive, and yellow, many with strong eyebrow lines and eye rings. The best way to distinguish them from the warbler is to watch their behavior. The vireo moves more slowly and patiently in search of its insects.

ABOVE: Vireo silhouette

- **Size:** Small
- **Shape:** Slightly larger than a warbler, with a heavier, hooked bill
- **Attitude:** Movements are deliberate—sits still and looks over a location before moving on
- **Behavior:** More solitary than the warbler; will travel in smaller mixed flocks
- **Common family members:** Vireos

Many other common families of birds may live near you. Woodpeckers, hummingbirds, owls, and hawks all have family traits that you may want to get to know better.

ABOVE: A cute little bird, but what kind? This Philadelphia vireo is very similar to both the warbling vireo and the Tennessee warbler.

"'Hear! Hear!'screamed the jay from a neighboring tree, where I had heard a tittering for some time. **'Winter** has a concentrated and nutty kernel, if you know where to look for it.'**"**

—HENRY DAVID THOREAU

Backyard Bird Feeding

The fun is about to begin! The very best way to start on your journey toward understanding birds is to have the birds come to you. Now is the time when learning what they like to eat, how they travel, and where they live provides huge dividends for you.

How will the birds hear the news about your feeder? It is probable that your first visitors will be one of your small, local, year-round birds. Chickadees, nuthatches, and titmice are always on the lookout for a new food source, and will probably find you first. It may take a few weeks for more birds to find your feeders. After that, you may see the floodgates open! Other birds will notice the activity and be drawn to check it out.

If you notice that interesting birds are stopping by and not staying, take time to learn about what they like to eat. Some of our favorites, like cardinals and juncos, may prefer to eat on the ground. Finches and pine siskins are on the lookout for weed seed. Woodpeckers don't have a lot of interest in seeds, but they will definitely become regular visitors to your suet feeder. Once your feeder program is in full swing, you won't have any problems attracting customers.

LEFT: A familiar face at feeders across North America, the chickadee's affable nature makes him a welcome part of the mixed flock.

Bird Food Basics

Bringing birds to your yard is really the best way to get to know them. Feeding birds is a big industry, so you may feel a little overwhelmed the first time you visit your local feed store, nursery, or home center. Along with a wide array of decorative and specialty feeders, there are a lot of seeds, nuts, and commercial birdseed mixes on the market. All those choices! If you are a natural-born shopper, it may look like paradise, but it's not really necessary to spend a lot of money to start your feeding program.

With all of those seeds, how do you choose? Here are the basics:

Black-Oil Sunflower Seeds

These seeds are the "meat and potatoes" of your bird feeder. A little different than the traditional striped sunflower seed that might first come to mind, the black-oil sunflower seed has a higher fat content and a bigger meat-to-shell ratio that makes it very attractive to birds. Jays, cardinals, woodpeckers, and grosbeaks all love them, but their softer shell also makes them appealing to small birds like chickadees, nuthatches, and titmice. The good news is that these seeds won't sprout beneath the feeder. The bad news is that they are so popular, there will be lots of empty shells to clean up!

Hulled Sunflower Seeds

These seeds are commonly packaged as sunflower hearts, pieces, or chips. Everyone at your feeder will love them, but these seeds are an especially thoughtful choice for redpolls, Carolina wrens, and goldfinches that may have a hard time breaking into shells. Because they are readily eaten and have low waste, they are great for places where you don't want to deal with the mess left by discarded hulls. More expensive than seeds in the shell, you will find that a little actually goes a long way. Don't put out too many at a time, though; they need to be protected from the elements and will rot quickly if they get wet.

White Proso Millet

Sparrows, juncos, mourning doves, and towhees are just some of the birds that enjoy this seed. It is best scattered on the ground, where these birds naturally forage. A common component of birdseed blends, you may find that a lot of the regulars that come to your feeder push it aside in favor of the sunflower seeds, so if you want to attract birds that prefer millet, buy it separately and use it specifically for them. A word of warning: cowbirds, blackbirds, and other annoying

---

Starting Your Feeding Program

Not sure where to start? Your first shopping list should include:

- **one simple tube feeder**
- **one bag of black-oil sunflower seeds**
- **one storage container with a tight lid**
- **one suet cage feeder and a cake of suet**

Let the fun begin!

LEFT: Dark-eyed juncos are ground feeders that enjoy the millet often cast aside by other feeding songbirds.

birds like house sparrows also like it, so if you don't want to attract them, stick to sunflower seeds. Also, millet can sprout beneath the feeder, so if this is a concern, you may want to sterilize the seed before using it. To do this, place one pound of seed in a brown paper bag and cook on high in the microwave for five minutes.

Safflower Seeds

Safflower seeds are a favorite of cardinals, and some other species eat them as well. The main reason to use safflower, though, is because of what *doesn't* like it. Feeding safflower may discourage squirrels and a few nuisance birds, like grackles and starlings. This is not a sure thing, however; I have friends who say their squirrels got over their aversion and now go after the safflower feeders as well!

Nyjer Seed

Also known as thistle seed, this food is essential for attracting finches and pine siskins, especially during migration. These tiny seeds are heat-sterilized so you don't have to worry about sprouting weeds underneath your feeder. They blow away easily and are kind of pricey, so tube feeders with tiny openings are made especially for this type of food. Nyjer seed needs to be fresh, so place it where it can stay dry and keep an eye on it for signs of mold. Give the feeder a quick shake once in a while. If the seed is clumping, throw it out and start fresh.

Nyjer seed can also be offered in "thistle socks." These fine nylon sacks are relatively inexpensive, allow air to circulate around the seed, and birds like to perch on them.

Peanuts

Who doesn't love peanuts? Peanuts are gaining in popularity as a bird food, and a number of species love them. There are even feeders designed especially for dispensing shelled and chopped peanuts. Jays enjoy whole peanuts in the shell, as long as they can get to them before your squirrels do! Peanuts are not as stable as sunflower seeds, though, so you have to make sure they stay fresh. Immediately throw out any that are darkening, as this is a sign that they are going rancid. Also watch for signs of mold.

Corn

Dried corn, either whole-kernel or cracked, is useful if you are trying to attract wild fowl like turkeys, geese, quail, or ring-necked pheasant to your backyard. Blackbirds, cowbirds, and sparrows will also eat corn, but so will deer, raccoons, and fox.

Commercial Seed Mixes

You know how your kids sort through the trail mix to get at the M&Ms? Birds are kind of like that when it comes to seed mixes. There are a lot of blends on the market, and good-quality mixes can attract finches, sparrows, and songbirds. Blends can be fine, but skip the bargain-basement blends. They contain a lot of fillers that birds just don't care about, like corn, oats, milo, and other cereal fillers. Look for sunflower-rich blends to make your birds happy, and don't be surprised if certain birds pick through the feeder looking for the pieces they want.

Suet

After black-oil sunflower seeds, suet is the most important offering you can make in your backyard. It provides a vital high-energy fat food source that is especially useful for birds,

RIGHT: The eastern bluebird is not tempted by your seed feeder but can be lured in with an offering of mealworms.

not only in the winter when other sources are scarce, but throughout the year. Suet attracts woodpeckers, nuthatches, chickadees, and blue jays; and wrens, creepers, and even cardinals might stop by for a quick snack. Pure suet should only be put out when the temperatures are cool enough to keep it from melting or going bad. Not only is melting suet messy in your yard, but the greasy suet can damage belly feathers. If an adult brings that gooey mess back to the nest, it can damage eggs.

During the warmer months, there are a number of warm-weather "no-melt" alternatives that contain corn, seed, and other ingredients. You can buy commercial suet cakes or make your own.

Mealworms

Mealworms are actually the larvae stage of a small beetle and are "bird candy" to a number of species. Some insect-eating birds that aren't attracted to your traditional feeders may be willing to stop by for a quick meal of fresh, live mealworms. I use them to bring in eastern bluebirds, although once the word has spread, everyone may want to stop by! During nesting season, even seed-eating birds will appreciate this high-energy food source. They are expensive, though, so keep them for those birds that you really want to have them, and put out only a few at a time. If you don't like the idea of dealing with live mealworms, there are freeze-dried or roasted versions. Birds don't find them very tempting because they are attracted to the wriggling of the live ones. Some people try to make them more appealing by reconstituting them in olive oil or water. I don't think this method is particularly effective. I do use dried mealworms in bird food recipes or as a topping to a suet recipe.

Feeding Programs by Season

The life cycle and migration patterns of birds create different needs for each season. Early spring arrivals have different requests than your year-round birds. Weed-seed eaters drop by later and want weed seeds. Winter birds are still looking for some of their summer and fall favorites. Watching and learning about the needs and lifestyle of your birds will help you to learn how to supplement their diets appropriately, and help you to attract visitors that you would like to see.

Follow this plan for a healthy bird-feeding experience year-round.

Autumn Feeding

If you are just getting started, this is the season to begin. Migrating birds appreciate the convenience of a quick, high-fat meal as they travel south, and year-round residents are happy to find that you will be a source of their winter's meals. Start with basics like black-oil sunflower seeds, peanuts, and suet cakes.

If you already have a feeding program, now is a great time to do a review. After a long summer of relative inactivity, feeding season is about to start in earnest. Check all of your feeders and replace any broken ones. Add any other feeders that you want to use this season.

Offer hummingbird nectar throughout the fall for migrating birds. As flower supplies dwindle, nectar feeders become more attractive, so I add extra feeders starting in August. In Virginia, my feeders stay up until November to make sure I have a meal available to migrating birds, but when to take them down varies regionally. When you take yours down, clean and sterilize it before storing it for the following season.

I put more seed feeders up in anticipation of heavier traffic. In September, I deep-clean all feeders that have been up throughout the summer, and I re-clean feeders that have been in storage. If I have had suet feeders up, I do a thorough soak in a grease-busting detergent. The holes in my log feeders get a spray and scrub before refilling and hanging them. Wooden platform feeders need a soapy scrub, too. Put your wooden feeders in the sun to dry thoroughly before you put them back out.

Take a look at your feed supply and discard any seeds that are looking old. Sunflower seeds in the shell are good for a year. Nyjer seeds are good for only six months or so. Suet and suet cakes should be stored in the freezer and will keep for up to a year. I always freshen up my supply of dried fruits and shelled nuts at this time.

Winter Feeding

Now is the time when your birds are counting on you to come through. Maintain a rich assortment of seeds, white proso millet, and suet at all times. High-calorie treats are very welcome at this time of year, so think fat! Suet, peanut butter, and lard mixtures will all be popular. Fruits, both fresh and dried, are also much appreciated. I also offer more nuts in the winter.

Winter is an especially rewarding time at the feeder, because birds tend to congregate in flocks to find food. Once the migrants have moved on, your visitors will probably settle into a familiar group of mixed birds. In cold weather, and without much natural food available, they seem to linger longer at the feeder. This gives you the opportunity to study some of your favorites a little more closely. The chickadees, nuthatches, and titmice come together, but they are not averse to allowing other visitors to join them at the feeder. Woodpeckers and juncos may also be

RIGHT: This gentle mourning dove perches patiently, waiting for its chance at the feeder.

ABOVE: Sudden winter storms put extra stress on birds who especially appreciate a chance to stock up at your feeder. These mourning doves and cardinals are waiting for their chance at the seed.

regular visitors. Depending on your region, you will have other regulars. This is a great time to learn their names and study their habits.

Keeping your feeders clean is really important now. Make sure that there is nothing in your feeding program that can cause the spread of illness. Dump seed that is old, and if you feed fruit or bready foods that can mold, make sure you clean them up before they make someone sick. I usually do my deep-clean of feeders in late December. Because I am filling feeders more often, I clean one feeder at a time, so there is always a food source in place. Now is a good time to rake up below your feeders before the snow gets too deep.

If you do notice a sick bird, it may be a good idea to take in feeders for a little while. Although you might really want to help the sick bird, it doesn't make sense to have it in there feeding with all the others and contaminating the feeding area.

Springtime Feeding

In spite of the burgeoning food supply in the backyard and forest, spring is no time to stop providing food. Your year-round residents are still stopping by for their sunflower seeds, cracked corn, millet, and suet. And guests traveling through will be thrilled to find a meal waiting, providing you with the pleasure of their company for a few days—or even a few weeks—before moving on to their next location.

Spring is an exciting time for seeing migrants. Depending on where you live, you should see flocks of finches, warblers, and sparrows sometime in April through early May. Take

The blue jay is a year-round resident, and a familiar face at the feeder in every season.

time to watch for their arrival and enjoy them while they are there, because who knows how long they are going to stay! Because they migrate at night, you may wake up one morning to find an empty yard and an abandoned feeder. This time of year, I have a friend who often calls in to work with "warbler flu." She can't bear to miss the passing flocks!

Finches are a delight at the feeder, especially in the spring when they are decked out in their best colors. Members of the finch family prefer tiny nyjer seeds, so a tube or feeder sock is the best way to attract them. It is important to keep this tiny seed fresh and dry, so check your feeder often and discard what hasn't been eaten within a month.

Even though you might not be able to identify which one is which right away, you will know when the warblers arrive. If the yard suddenly has a frantic, busy air that it didn't seem to have before, the warblers may have arrived. And those flashes of color that you didn't see before are a telltale sign. Warblers are not big feeder customers, since their primary diet consists of insects, but they can be attracted by mealworms and suet.

Winter Vacation?

If you are a winter bird yourself and plan to take an extended vacation to other climes, keep in mind the responsibility you have to the birds that depend on you. If you are in a neighborhood with other bird-feeding friends, or in an area that has some natural resources for your birds, a three-week vacation is not a big deal. But if you are in a cold climate, are the only local bird restaurant, or a winter storm comes up while you are away, a sudden lack of food can be dangerous for the bird residents that have come to depend on this food source.

Wicked Weather and Late-Season Storms

The problem with spring is that the weather is often unpredictable. Birds that made the decision to move north sometimes find themselves caught in foul weather or a late-season storm that wreak havoc on their food supply. Be prepared to supplement their needs with a special variety of foods for times like this.

If a storm hits, reach for these ingredients from your pantry:

- Peanut butter spread on tree trunks or pinecones (a great high-calorie snack)
- A crumbly mix of cornmeal, dried fruit, and peanut butter for bluebirds and thrushes
- Crushed nuts, such as walnuts, pecans, or peanuts, for small birds like wrens and catbirds
- Larger nuts for nuthatches and jays
- Oranges, bananas, and apples for orioles and thrashers
- Grape jelly for orioles
- In a pinch, a tray of moistened high-protein dry dog food will attract bluebirds and starlings.

If you are concerned about a late-season storm, have suet and mealworms available. Even if they are not a part of your regular feed, having them on hand can keep your feathered friends from starving or freezing to death.

Summertime Feeders

Good weather is bringing out the insects, fruit and berries are ripening, and the nesting season is on. The birds that you have been accustomed to seeing have paired up and gone off to set up housekeeping together. If you have made your yard into an attractive habitat, you may still have lots of birds around, but summer is not primetime feeder season, so I reduce the number of feeders I keep out, and stick to the basics. Sunflowers and millet are my primary menu items, and I generally don't have a lot of takers. Still, there are a few regulars who like to stop by and grab a quick seed or two for the road.

Suet can be messy in the summer, especially in the heat of Virginia. Melting suet stains decks and sidewalks. Birds do appreciate the opportunity to stop by for a quick high-fat snack, so I switch to no-melt suet cakes that resist melting and still appeal to most of my suet customers.

The natural food supply is abundant right now, and nesting birds have to make the most of that. There are several mouths to feed back at the nest, and, for most baby birds, insects are their food of choice. Even birds that normally eat seeds will be out in search of insects for their nestlings. Because it takes so much energy to raise their young, adults may appreciate the chance to stop by for a few seeds for themselves, and may reward you by bringing out their family to visit you and your feeder on one of their early outings. Fledglings enjoy soft treats such as suet and peanut butter while still getting the hang of dealing with seeds and nuts.

Recipes and Blends for Your Feathered Friends

For the most part, birds are perfectly happy with the basics. But there are some special treats that may attract hard-to-entice species or provide an extra boost for them during nesting and the cold season. We are especially mindful of our birds during the holiday season. Just as we are sharing our abundance with loved ones, we like to remember the little creatures outside our door who are just beginning a long, difficult season.

It is helpful to know what kinds of birds like what types of foods. My mother used to throw out scraps of bread, knowing that someone would eat them, and in fact, a lot of birds enjoy bread products. But others have no interest whatsoever. Keep the following list in mind when offering food from your own cupboard:

- **Bread and grains:** A lot of birds like bread, cornmeal, millet, and oats. Blue jays and crows are always ready to grab for a meal of bread. And other birds are perfectly content to have bread crumbs, oats, or cornmeal mixed into their food. Wrens, mockingbirds, thrashers, sparrows, warblers, titmice, cardinals, grosbeaks, buntings, and chickadees will stop by for this treat.
- **Fat:** Fats are a crucial source of fuel during nesting and in winter weather. Suet is the first ingredient that comes to mind when we think of birds. Suet is made from raw beef fat. It is a favorite of nuthatches, woodpeckers, wrens, titmice, chickadees, and cardinals. Even bluebirds will occasionally stop by for a taste. I keep both lard and suet around for making mixes, and stir in leftover bacon grease for a special treat. Other oils, like olive and corn oil, can also be added to mixes.
- **Peanuts and peanut butter:** Many birds share our enjoyment of peanut butter, but a lot of the brands that humans like are full of sugar and additives, so I keep a jar of natural-style peanut butter just for them. I use peanut butter in a number of my mixes. Just about every bird that likes nuts, seeds, and fat will come for a taste of peanut butter. Chopped peanuts have become more accessible, and even cardinals and finches like the taste.
- **Dried fruits (i.e., dried apples, raisins, and currants):** Dried fruits are the perfect replacement for birds who like to forage for berries during the regular season, including waxwings, orioles, and thrushes.
- **Sunflower seeds:** The sunflower seeds in your cupboard are probably the old-fashioned striped sunflower seed. But black-oil sunflower seeds have become the premium food of the bird world. They are high in fat, and wild birds from jays to finches enjoy the taste. The shell is slightly thinner than the old-fashioned striped sunflower seeds, and they have larger

kernels. So while your birds won't mind a taste of your sunflower seeds, they would probably prefer the black-oil seeds that you keep just for them.

- **Apples:** Apples are an easy-to-keep fall and winter fruit for us, so it's pretty easy to offer fresh apples in mixes, or cut up and left at the feeder. Because birds eat early and late, it isn't necessary to leave fresh fruit out all day. Put it out at breakfast and again at dinnertime, and refrigerate it in between. A lot of birds enjoy fresh fruit in the winter, including waxwings, mockingbirds, thrashers, wrens, cardinals, grosbeaks, and buntings.
- **Oranges:** To me, a fresh-cut orange left on the feeder seems like one of the most hospitable sights of winter, and a number of birds will come for a taste, including orioles, finches, tanagers, and even woodpeckers. Place half an orange onto a large nail that is hammered into a stump or fence post, or position them on platform feeders or even in suet cages.

Bird-Friendly Habitats

Spring is a busy time for birds. They are traveling from their winter grounds, going through their prenuptial molt, and getting ready for mating season. They are starting to search for nesting areas and preparing for their courting rituals. As the weather warms up, they may start to abandon your feeder in favor of fresh insects and new nesting territories.

Most birds are looking for a good source of insects in the spring, so watch for your birds to be on the hunt. Most birds that have been loyal to your feeder all winter are anxious to get out and find some fresh protein. Caterpillars are a favorite of your cardinals. Bluebirds and towhees love beetles, grubs, and spiders. Warmer weather brings our first outbreaks of mosquitoes and other flying insects, which will put warblers, phoebes, flycatchers, and cedar waxwings on the prowl. As soon as the ground warms up, you will quickly note the arrival of robins in search of earthworms. The burst of new green leaves brings out other insects on branches and in bark. Look for busy chickadees, titmice, nuthatches, and woodpeckers to be busy there. Open water offers up larvae and mayfly hatches for a wide range of birds. Don't like bugs yourself? Never fear! A healthy yard full of insects always attracts birds that are glad to keep the place safe for you.

Phase One: Foundation Planting

While your birds are out hunting, now is the time to do a little planting. My plan for spring planting includes two phases, and phase one is foundation planting.

Where in the yard am I going to invest in a new tree, some new shrubs, or some perennial berry bushes? Budgetary considerations require me to make just a few such purchases a year, so I take inventory of my yard to decide what I want to add. Usually my husband and I have dreamed up projects over the winter and know exactly what we want to do, so it's just a matter of prioritizing. But if you are new to wildscaping, consider the following:

Hiding places: Your birds need natural shelter, so think about creating more hiding places for them by adding berry bushes and flowering ornamentals. Look at travel routes around your yard. Is it easy for birds to travel under cover, or are there wide-open spans of lawn? A mixed hedge of shrubs makes for easy travel, as does the addition of some shorter trees near tall ones. Shrubs are also ideal safe nesting sites for wrens and other small birds that like lower locations. Choose barberry, hawthorn, lilac, privet, or any number of a wide array of berry-producing bushes.

Shrubs: If you were planning to buy shrubs anyway, look for natives with food-producing capabilities. It is not hard to please both yourself and the birds. The amazing burning bush produces a fantastic pop of color that I love, and the birds really go for the bright berries that come along with it. Beautiful evergreen hollies are native to our area, and the birds love the red berries they produce late in the season. But don't think a single specimen bush is going to make the ideal bird home. A lone shrub in the middle of the yard says "Look here!" to predators. Variety is the key. Try groupings of three to five shrubs together to make a pleasing corner of the yard that is also bird-friendly.

Layers: Think in layers; find ways to fill the vertical space between that tall oak and your lawn. Some small trees, like dogwood, adapt naturally to that shady middle space. Shrubs add the next layer down and perfectly suit small birds that like to nest only 3 to 5 feet off the ground.

Go large: Tall trees such as ash and tulip provide plenty of seed. Oaks are great for acorns. Large evergreen trees such as juniper, spruce, hemlock, and fir offer great shelter in bad weather and provide food for cone-seed-loving birds.

LEFT: Even the more-domesticated parts of our yard provide a safe travel route for our birds. Although not strictly a native, the dwarf Japanese maple in the foreground was in place when we arrived, and at thirty years old, has earned its place in our yard. Evening grosbeaks enjoy the seeds.

Consider the middle ground: Shorter, decorative trees like dogwoods and smaller evergreens grow in the shade of taller trees and produce plentiful fruit in the fall. Fruit trees such as cherry, Juneberry, mulberry, and crab apple are always popular with the birds. Consider planting a hedgerow of mixed tall shrubs with fruits and seeds that mature at different times. Sumac and dogwood are a great source of fall food; rose bushes offer berries long into the late seasons; and holly and hawthorn provide berries in the winter months.

Low and lovely: Berry shrubs (like blackberry, raspberry, blueberry, grape, and strawberry) and evergreen shrubbery (like rhododendron, yew, and spirea) are great for nesting birds. Berry-bearing autumn shrubs like winterberry, spicebush, sumac, and Virginia creeper are welcome after the summer fruit is gone. Hummingbirds love azalea, butterfly bush, and weigela, among other flowering shrubs. Shrubs that offer berries into the winter include hollies, barberry, sumac, and roses.

Ground play: Don't forget the bottom layer of your yard. For a partially sunny area where no one walks, consider adding vines. A tangle of honeysuckle will thrill the hummingbirds, and a stand of Virginia creeper will attract all manner of birds. Bittersweet and wild grapes are also great choices. And that crop of poison ivy? Birds actually love the white berries they produce, so if you find an out-of-the-way patch, you may want to note that area for bird-watching!

Phase Two: Flowers and Vegetables

Phase two of spring planting is flowers and vegetables. Keep your birds in mind when you choose your flowering plants.

Flowering plants: A garden filled in springtime with red and orange nectar-producing flowers will attract hummingbirds later in the summer. It will probably attract butterflies, too, which may in turn attract flycatchers and other birds seeking flying insects.

Think color: Choose red salvia, columbine, honeysuckle, coneflowers, and fuchsias for hummingbirds and other nectar lovers.

ABOVE: A stand of native grasses and vines softens the edges of our yard and provides seeds and berries to the local birds.

Annual investment: The annuals that you plant are not only good for providing spot color throughout your garden and planters, but their extended blooming time also provides seeds and nectar long into the season.

Go wild: Sparrows and finches make the most of wildflowers and weed seeds, so if you have room, consider a corner of your yard for assorted wildflowers. Some wildflowers that thrive include goldenrod, yarrow, coneflower, joe-pye weed, and black-eyed Susans.

Attract bugs: Plants like sedum and goldenrod attract insects, which in turn attracts insect-loving birds.

Building fund: Willow and milkweed produce food and nesting materials for weed-eating birds.

Cultivate late bloomers: Late-blooming flowers like salvia, hollyhock, and lobelia are greatly appreciated by hummingbirds as they start their fall migration.

Fruit and vegetable gardens: As to our own fruit and vegetable gardens, we are willing to share—to a degree. If you are trying to grow your own food, birds are a good news / bad news situation. Sure, they help a lot when it comes to natural pest control. They are happy to cruise your beds for insects that might be eating your leafy greens or chomping on your tomatoes. But they also have a bad habit of picking up seeds you just planted or helping them-selves to juicy strawberries. (Okay, I wouldn't mind sharing, but couldn't they just take the

ABOVE: Native flowers and grasses provide a wealth of weed seed to summer nesting residents.

whole berry rather than hopping along, putting holes in each one like the visiting blackbirds do?) To mitigate the problem, we have a few strategies:

1) **Skip the seeds:** We use bedding plants instead of seeds for as many of our vegetables as we can. That way we don't lose whole rows of plantings before they even get a chance to germinate. Start your seeds indoors to get a jump on the birds.

2) **Hire security guards:** Our dogs cruise the area where the vegetable garden is. They are pretty good at keeping marauding flocks of crows or blackbirds from making themselves too comfortable. And while they squabble a bit with the birds, they have never hurt one. If your dog is an aggressive bird hunter, you may not want to try this one.

3) **Provide alternative food:** We also grow a few sunflowers specifically for the birds. We keep a feeder nearby. I think we may be kidding ourselves here. After all, if you were given a choice of fresh fruit or the same seeds you have been eating all winter, which would you pick? But it makes us feel that we are trying to be fair.

4) **Create a barrier to entry:** We put bird netting over the parts of the garden that we really don't want to share. At least, we try to. Every May, when the cherries start to form, I say to my husband, "Time to put up the netting." I say this every day for a few weeks. Then one day, we each reach for the first bright, almost-ripe cherries. Delicious. Okay, that net has got to go up tomorrow! As soon as we make that resolution, however, the birds decide the berries are ready. Inevitably, they clean the tree of every piece of fruit by the very next day. No cherries for us . . . again. But we know what a thrill those trees are to our birds.

ABOVE: Security guards keep our vegetable garden safe from flocks of blackbirds.

5) **Raise a ruckus:** Scarecrows, shiny pie plates, or wind chimes can be used to ward off birds. Think shiny, reflective, and noisy. Flapping banners, streamers, or flags only work in the short term, as birds tend to get used to them. We gave up these methods long ago. Our birds just seem to find them amusing!

6) **Harvest promptly:** Birds have a good eye for fresh food, so if you want your fair share, try to take a walk through your garden and harvest every day. Overripe vegetables attract insects that bring more birds into the garden, and can lead to them investigating plants that wouldn't be interesting otherwise.

7) **Share the wealth:** Chop overripe melons or squash in half before throwing them on the mulch pile. Birds that like the seeds will be happy to find them. At the end of the season, leave some apples or berries behind for the birds, and let old crops stand for fall and winter foraging.

RIGHT: Late-season strawberry beds are fair game for the birds.

Donkeys

NO, THEY AREN'T STUBBORN. Donkeys, or long-ears, are highly intelligent, and like other highly intelligent beings, they don't just obey because they are told what to do. And if you end up being kicked by a donkey, you most likely deserved it (perhaps you called him or her an ass?). Donkeys are their own species. Male donkeys are called jacks. Female donkeys are called jennies. Mules are the product of breeding a male donkey with a female horse. The offspring of a female donkey bred to a male horse is called a hinny. Mules and hinnies are sterile and cannot reproduce (which is a good thing for a hobby farmer). Donkeys can live to be thirty years old or more. Donkeys are known to bray, which is the hee-haw sound for which they have a reputation. Ours never did this until we put our llamas in the field with them. Now they do it mostly just to show off when a visitor arrives at the farm.

Jacks are easy to come by and are less expensive than jennies. Since we don't advocate breeding, you won't have to be as picky about the characteristics when buying a donkey. Ugly donkeys need homes too. Most people recommend that you geld males to make them

less aggressive. But as long as you don't have any females around, there shouldn't be a problem, although every situation is different and jacks can be dangerous. Bring a female horse or donkey within a mile of an unfixed male and you'd better watch out and be very careful. That's another reason why you should consider staying away from breeding.

We've had two unfixed males that were raised together for eight years and never had a problem. They do wrestle quite a bit and mount each other (that's a domination

thing, not a sexual one). But all that activity keeps them in shape. Unless they are really injuring each other, being aggressive to people or other animals, or if they are show animals, there should be no reason to have them fixed. Gelding donkeys is inexpensive and easy though, so if you have any concerns or have children, go ahead and have a veterinarian do it. Donkeys are strong animals that can kill a person (well, maybe not a miniature donkey), so always err on the side of caution.

Donkeys are terrific guard animals, especially if you have sheep or goats. Those ears are long for a reason—so they can hear trouble coming from miles away. We once saw a herd of gazelle at a Texas ranch being watched over by an alert donkey, standing at attention, ears scanning the distance on a hill above where the herd was resting. Our donkeys chase any stray dogs out of their field. Donkeys provide manure that is ideal for compost production. Unlike horses and cows that just defecate wherever they're standing, donkeys will create neat piles. We just drive the tractor around the field about once a month and shovel the piles into our front-end loader, pile it up to age, and later add it to the garden or compost piles.

Donkeys are desert animals that originally came from Africa. On the (relatively) easy street of Hobby Farm, USA, they can easily become overweight and can founder on a field of lush grass. No one knows the real cause of founder, but it is a condition that happens to

horses and donkeys, typically when they are feeding on very rich grass after a rain, in which the bone of their foot pushes into their hoof wall. This causes extreme pain and lameness, and if not caught early, can mean the end of the animal. Fat donkeys and horses are more susceptible to foundering.

Feeding

Donkeys get very wide when overweight and the crests of their necks bulge and can even fall to one side under the weight. This should be avoided at all costs because it's mostly a non-reversible condition. We made the mistake of feeding our donkeys treats of alfalfa cubes when we first had them. We only fed them a small amount in a bucket each day. They loved it, but quickly became overweight and began to have hoof problems because of the excess protein they were receiving. As soon as we took them off feed and only allowed them to eat grass and hay in the winter, they dropped the weight and the

RIGHT: This donkey has put on a little too much weight and it's showing in the crest of his neck.

Estimating a Donkey's Weight

To estimate an average donkey's weight using a tape measure and calculator, measure the donkey's **heart girth** in inches, which is completely around the body of the donkey a couple of inches behind his front legs. Measure the **length** from the point of the shoulder to where the tail meets the body. Measure the **height** from the top of the donkey's withers (the middle of the cross on most donkeys' back) to the ground next to the hoof. Then use this formula: Weight in pounds = heart girth (inches) x height (inches) x length (inches) / 300. This is a rough estimate and if you really need to be exact because of medication (it's okay to estimate when you're just deworming them) or some other reason, you should have a veterinarian check the weight.

health of their hooves improved. Under the record drought we've had this year in Virginia, we've not fed the donkeys any supplemental feed and they've never looked healthier. You may need to resort to a grazing muzzle or to keeping them in a smaller pen to keep their weight in check. If you do need to feed them grain in the winter (if you run out of hay, for instance), then buy the lowest protein mix you can find, preferably 10 percent or lower. Also, they need a mineral lick. It's the big, pinkish block available at your local feed shop. Just tell the clerk you need a mineral lick for your donkeys and he or she will point you in the right direction.

The same plants that are toxic for horses are toxic for donkeys, although donkeys seem to be less susceptible to eating them in enough quantity to do much harm. You should contact your local agricultural extension agency to find out what native plants you might need to watch out for.

Maintenance

Some things you'll need to do to maintain your donkeys are to deworm them every six months, get them a tetanus shot once a year, and keep their hooves trimmed. There's some debate about whether donkeys need to be vaccinated against West Nile virus. We've never done that and our vet didn't believe it was necessary. The jury is still out on this, so you'll need to make your own decision. But donkeys don't seem to get diseases as frequently as horses and it's generally believed that they are more resistant to many of the ailments that affect horses.

To deworm a donkey, all you need to do is buy a horse deworming medication from your local feed or farm-supply store. Get the kind that you can squirt into their mouths. These also have a dial on them so you can adjust to the weight of the donkey on the applicator, insert it into the back of the donkey's mouth (go in from the side), and squeeze it in. You only need to do this about every six months.

Trimming their hooves is another task altogether. Donkeys are used to rocky ground that naturally keeps their hooves worn down. But most farms are all grass, so donkeys' hooves grow unimpeded. Many people have a farrier come out to trim hooves, but farriers are busy and don't like to deal with donkeys. Because donkeys don't need shoes like horses, there's no cost benefit to a farrier taking the time to come out to do it. But you should talk

ABOVE: Donkeys love a good roll in the dust.

one into coming at least once to show you how. Or have the person you bought them from show you. It took us several lessons to learn the proper way. Neglected, a donkey's hooves can grow out and curl up to resemble a genie's slippers. This is a very painful condition that requires serious veterinary help and may mean that the animal has to be put down. Hoof care is the most important task of the donkey owner. Make sure you learn it properly and keep the hooves trimmed at least every two months (every six weeks is preferable).

Donkeys can also be used for work. They can be harnessed to a cart, ridden by smaller people or children, or used to pull logs from the woods. This is something that you might explore after getting the basics of care down. But banish any ideas that you might harbor of making any money from donkeys. You might get a few bucks here and there from petting-zoo type work, but manure is the most valuable product that a donkey produces. That and protection for your other animals.

Calling a Donkey

Call, "Heeeeeee-DONK! Heeeeee-DONK!" Shake a bucket of treats in between your calls and you'll get them every time.

You should visit the American Donkey and Mule Society (ADMS) at www.lovelongears.com before buying a donkey or mule and even after. They have the best resources for long-ear lovers.

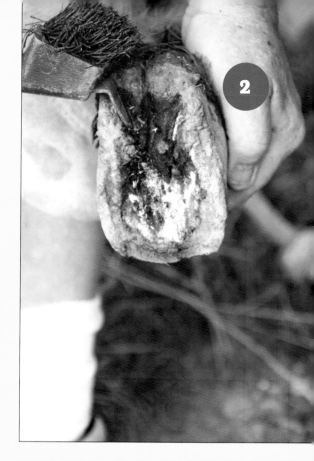

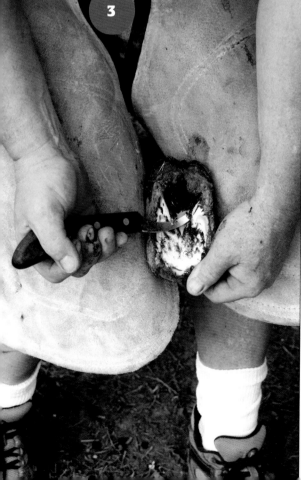

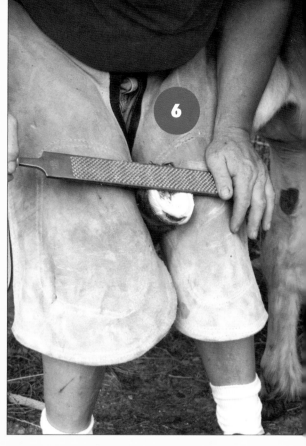

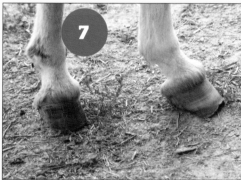

Trimming a Donkey's Hooves

1. Hoof pick, file, hoof knife, and nippers.
2. Dig out manure and mud with hoof pick and brush.
3. Use hoof knife to cut away access frog (soft part of the hoof that comes to a point in the middle) material.
4. Hoof and frog should look like this when done cleaning with the knife.
5. Trim off excess hoof material all the way around the hoof. If you draw blood, back off and don't cut so much off.
6. File the hoof flat and take off any sharp material.
7. The hoof on the left has been trimmed. The hoof on the right needs it.
8. Ride your donkey into the sunset.

Do not attempt this on your own until you've had an experienced farrier come to your farm and demonstrate for you. This is the basic technique, but there are many variables and nuances that need to be demonstrated in person.

Handling Large Animals

Temple Grandin, in her book *Animals in Translation*, describes going out into a field of skittish cows and just lying down. The cows, being naturally curious animals, soon gathered around her and quietly checked her out. Putting herself down at the level where they keep their faces most of the day was how she connected with them and convinced them she was no threat. Cesar Milan, in his show *The Dog Whisperer*, advocates a calm, assertive demeanor around animals. We take both approaches.

Animals like horses, cows, longhorns, donkeys, and llamas are powerful creatures that can kill or maim a person if startled or threatened in the slightest. They can sense fear too, like all animals. So always screw up your courage when approaching them, control your breathing, and be forceful without being cruel. Talk to them in low, reassuring tones. They don't speak baby talk, so avoid that temptation. Avoid making sudden moves around them. When walking behind them, give them a wide berth, or keep your hand running along their body so they know where you are. And once you're comfortable, then get down on their level like Temple describes in her book. You'll be fascinated and surprised at the perspective you gain looking up at them instead of the usual standing posi-

ABOVE: Rico the llama peeking out from the llama shed, where he's free to come and go as he pleases.

tion that has you always looking down on them. But if an animal is acting out, it's okay to give him a swat on the hindquarters to let him know the behavior isn't appropriate. These are big, tough animals and can take a firm smack on the rear. If you have a very unruly animal that you're having a hard time caring for, you should call in another more experienced farmer or veterinarian to help you find strategies for dealing with the animal. And there's one almost foolproof way to get an animal to do almost anything—find his favorite treat, like apple and oat treats or sweet feed shaken in a can.

Shelter

Large grazing animals require little in the way of shelter. In most climates, a three-sided shelter with a roof is about all that you need to shade them from the sun or protect them from the cold, and much of the time the animals won't even use it. If you live somewhere that has frequent or even occasional heavy snowfall, then you will want to offer a shelter where animals can dry off. It's usually not the cold that's the problem, but the wet. If you plan to

breed your animals, you'll definitely need a place for the newborn and mother to shelter until the newborn is ready to be put out into the field. And you'll need to plan for separate living quarters for your unfixed males.

We have a small stable for our two donkeys and two llamas. They share it and tolerate each other. It has no door, so the animals come and go as they please. It has two windows, which we open in the summer and close up for the winter. We also have an area to store hay where the animals can't get to it. The more time they spend in the stable, the more manure builds up that you'll need to muck out. When we had over fifty inches of snow this winter, the animals spent a good deal of the day in the stable, but they would come out to eat by our house when we'd call them. The llamas could have lived outside the entire winter, had we not spoiled them! They are mountain animals, after all.

Cows and longhorns are even tougher and typically prefer to shelter under trees and not in enclosed areas. You can ask local farmers what, if any, shelter you might need for your climate. Remember that cows live out on vast ranges in the West and Rocky Mountains and they're not given shelter. Of course, they freeze to death in these places occasionally, but that's certainly not common.

The best fencing for donkeys, llamas, and horses is board fencing or woven or high-tension wire fencing. Some people use electric fencing, but large animals will eventually break through it or the deer will inadvertently knock it down. Electric fencing is ideal if you are employing rotational grazing. Barbed wire is strictly discouraged if you have any animals other than cattle. Cattle are bigger, tougher, and more destructive and most need to be kept in check with electric fencing or barbed wire.

Food

Large, grazing animals can live on grass alone, with minor mineral amendments. That is, if you have enough for them. The more feed you have to give them from off the farm, the less sustainable and economical is your farm system. We keep our animals mostly for the compost we create from their manure. But if we had to spend $1,000 a year feeding them in order to get that manure, we'd be getting no benefit from them outside of their companionship (which is a lot in our opinion). One animal per acre is a traditional rule of thumb. But we've found that old rules of thumb, especially when it comes to raising animals, can be unreliable. Every situation is different.

You should supplement grass with minerals. Cows can typically get by in most places with only a salt lick. Llamas, donkeys, and horses should be provided with a mineral block to which they can have free access as they see fit. Horses should always have access to salt as well. But certain areas of the country are deficient in certain minerals. And if your pasture had been poorly managed before you bought it, then it may have deficiencies that you would need to supplement. This is exactly what your agricultural extension agent is for. He or she can help you figure out what nutritional needs, if any, aren't being met by your pasture. For our llamas and donkeys, we leave out one mineral lick and one salt lick at all times.

If you want to keep your fields healthy, three acres or more per animal is a safe bet. If you have enough pasture and only a few animals, you don't even need to rotate them and they can forage at will, never stressing the pasture. But if you do need to rotate, say if you have eight animals and ten acres, then you'll want to split your pasture into paddocks (using movable electric fencing is a good way) and rotate your animals every few months. It's ideal, if you're rotating, that your animals are never on the same piece of pasture twice in one year, so take that into consideration when you decide how many animals you're going to raise on the land you've got. If you've got a chicken tractor, it's a good idea for it to follow the cows wherever they've been last so as to spread their manure and till it into the ground. Everywhere your chickens go will end up having the best grass you've got.

You'll need hay for the winter. It's tempting to make your own, but haymaking takes a lot of expensive equipment and a whole lot of physical labor. So buying is probably your best option for at least the first several years you're getting started on a farm. Plan ahead though, as hay can become quite scarce at many times of the year and when the weather has affected the annual crop. It's best to buy early in the summer, after the first cutting. You never know if drought will set in and there won't be a second cutting.

First-Aid Kit

The best thing to have in any first-aid kit is the number to your regular veterinarian. You should have the vet out immediately after you bring any animal to your farm to establish the relationship and have your animal checked out, much like you have a mechanic check out a car you're buying.

Every first-aid kit is different. It's not a bad idea to buy one that is already put together. You can find these at your local feed store or online. Trying to round up all the ingredients (gauze pads, wound dressing, hydrogen peroxide, iodine solution, animal thermometer, etc.) yourself is tough. But an animal first-aid kit isn't much different than a human one, which you can certainly use in a pinch.

Dealing with Death

Tom Dance, of Oak Creek Ranch in Fredericksburg, Texas, has become the undertaker of sorts for the animals from neighbors' farms. He's the only farmer within the immediate area who has a backhoe on his tractor. So whenever a horse or cow meets its timely, or untimely, end the neighbors call Tom to come dig the hole, push the animal in, and cover it up. When it's a particularly beloved animal, like a horse, the neighbors sometimes gather together and hold a funeral of sorts, saying some words of appreciation and prayer. There's sometimes a party afterwards too, just like a human wake.

A backhoe is the preferred method of taking care of a deceased large animal. If you don't know of someone with a backhoe, your vet probably does. They'll trailer it over for a fee and bury your animal. If you have to put an animal down, always try to have a vet do it. If the vet is unavailable and the task is left to you and any friends you know that can help, then you'll need to put a .22 caliber or larger bullet right in the forehead of the animal. Don't hold the barrel directly against the forehead. Be careful and quick about it.

For cows, the most common method over the years to dispose of their bodies after death is to burn them. You might not be able to dig a hole big enough in rocky ground to bury such a large animal. In fact, burning is really the best way to deal with the body of any animal as it keeps other animals from disturbing the carcass and spreading disease. It also does away with the inevitable rot and smell that might come with a shallow burial. All you need to do is pile a bunch of brush and logs on top of the animal, essentially a funeral pyre. You'd not think a big animal full of liquid would burn, but if you add enough wood, it'll burn eventually. You might also check first with your local health department to make sure this is legal in your area.

⑥

Llamas and Alpacas

OUR LLAMAS ALWAYS GREET us first with their soft noses. They stretch their necks out, we reciprocate, and our noses come to within centimeters. We puff some air out at each other, and if we've brought the treats they are looking for, we hold them in our mouths and the llamas purse their big, soft lips and gently take them, chewing with the side-to-side motion you know from seeing camels in movies or at a zoo. Their lips are so pouty and soft; it's hard to believe they spend all day using them to gather grasses as deftly as if they were human fingers.

Llamas are very birdlike in their movements. Their slender legs and small, clawed feet seem almost too small for their big bodies and necks. But llamas' size is due mostly to their woolly fleeces, and after a shearing they reveal themselves to be quite thin and more squarely proportional. They step lightly and carefully. They appear gangly like a pubescent teenager when they are in a slow gallop, their small heads bobbing up and down every few steps. But there's nothing more graceful than a llama in a full lope. They don't run, they spring, all four feet leaving the ground together, bouncing along, heads held high, bringing to mind Tigger or Pepé Le Pew.

Llamas are more effective guard animals than our much larger and more powerful donkeys. They rarely make a sound, but when they are alarmed they emit a high call, so unique it's impossible to describe properly, but it's similar to some turkey calls. We have two llamas, and when a predator or dog enters our field, the llamas quickly come together, teaming up shoulder-to-shoulder to maximize their size. They then walk together toward the intruder and lower their heads to the ground as they walk. It's quite an intimidating show and there's not an animal yet that's ventured to test their mettle.

Llamas and alpacas are cousins. Domesticated by the Incas, they are part of the animal family of camelids. Llamas were bred mainly for use as work animals while alpacas were bred for their fiber. Alpaca wool is much finer, easier for spinners to work with, and more valuable than llama wool, in most cases. You care for each animal in pretty much the same way, taking into account that alpacas are much smaller and less able to defend themselves.

ABOVE: Ferdinand and Rico come running for their favorite apple and oat treats.

If you are going to raise alpacas, you should have a llama around to protect them. We'll only refer to llamas from here on out.

Llamas are ruminants, like cattle. But unlike cattle, they have three stomachs instead of four. Ruminants chew their cud. Chewing the cud is how ruminants digest their food. They eat it initially and it goes into their first stomach to partially digest. It's then regurgitated and they chew it again. It's the digestive juices in the first stomach that become the infamous spit that ends up on offending creatures and people. It smells about like you'd imagine partially digested food to smell. Llamas have bottom teeth and a hard palate at the top of their mouth against which they grind their food. They do have two upper pairs and one lower pair of sharp, fighting teeth. Males use these teeth to fight for the affections of females.

Llamas' reputation for spitting is probably very much a part of their evolutionary plan. We've had two gelded male llamas for five years and we've never once been spit on, but that's probably because they have no females around to arouse them. They do spit at each other occasionally, but only when they are competing for their favorite apple and oat treats. Llamas are large, imposing animals, but they rely on intimidation mostly to deter predators. They have dainty legs that, while powerful, don't carry the punch of a horse or donkey hoof. So the legend of the spit helps them keep most people at a distance.

Maintenance

There are two regular tasks that you'll need to learn to do for your llamas unless you want to pay someone to come out every six weeks to do it for you. And you should have someone show you at least once before attempting yourself. The first is to deworm the llamas.

Audrey administers the de-worming medication by injecting it into Ferdi's shoulder.

You'll also need to learn to trim their toenails. This is easy if you can learn to grab the leg and secure it before they pull it away. You'll need to be very careful when doing the back legs as they typically will jerk their foot out of your hand and then kick you square in the behind.

It's recommended that you have your llamas and alpacas sheared once a year. This is especially important in hot climates. Llamas are mountain animals and are easily susceptible to heat exhaustion that can lead to death. If your llama is foaming at the mouth, it's overheating and you should take steps to cool it off, including running fans on it and shearing it immediately. If you are planning to sell the fleece, you'll need to learn from someone with experience and you'll need practice. The fleece needs to be trimmed in a certain way for it to be desirable to spinners. You should have someone with experience demonstrate for you in person.

Trimming Llama Nails

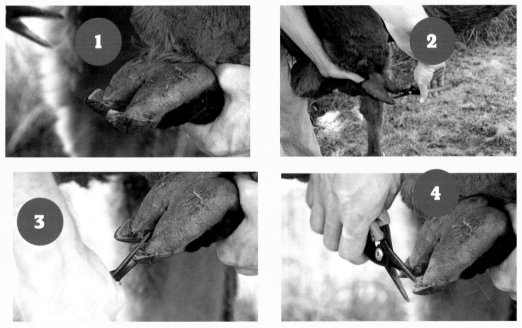

1. You can tell this llama's nails need trimming—they curl around below the pads of his foot.
2. Hold the llama's leg up and use straight clippers. Straight gardening shears work fine.
3. Trim the sides down.
4. Clip the tips. You want the nail to go straight, curve at the end and touch the ground flat with the rest of the foot.

Hand-shearing a Llama: Jonathan Sides demonstrates the proper way to hand-shear a llama. That's one skinny llama under all that hair!

Cattle

IN THE OLD DAYS every homestead with a family to raise had at least one cow on pasture or in the barn. To the farmer who kept her she'd supply a source of milk for nine to ten months of the year and enough beef to feed the family throughout the next year by her offspring.

Her milk was fed fresh at the table or made into cheeses, yogurt, ice cream, and treats, plus any surplus helped to sustain pets or her barnyard companions. The beef of her young, raised nearly without cost on pasture until the snow fell, was higher in vitamins and nutrients than today's grocery store beef, even with all of our modern-day advances in science and animal husbandry. The reason modern beef pales in comparison is all in how it is raised. Beef that shows up in your grocery store is more often than not the product of cattle finished on a diet of grain in a feedlot. A lot of grain in fact. And for many months.

On the surface this doesn't sound odd. After all, grain increases bulk and marbling within the muscle of cattle. With all that extra fat throughout, the beef is juicy and delicious. But the problem is that cattle digestive systems were not meant to survive solely, or even mostly, on grain. In fact, all that grain makes cattle sick, causes liver dysfunction, and stresses their immune systems.

Feedlots counteract cattle's reluctance to consume so much grain by giving them hormone shots that increase their appetites. Now the cattle will eat almost anything. You can research precisely what "almost anything" is online or through recent documentaries. I promise you'll never touch a grocery store steak or a fast-food burger again. The cattle, sickened by the feed they cannot help but eat, are then given medications to fight infections and override liver, organ, and system failures. Sicker still and loaded with chemicals, hormones, and antibiotics, the cattle's discomfort becomes an acceptable norm. Acceptable to the feedlot owners, perhaps, but not to the cattle.

One day, this has to stop. This is why I applaud every person on a mission to raise his or her own food. Not to upset the balance of corporations profiting from pain—after all, not all feedlots are created equal—but to reinforce your ideals with one more reason to carry through.

A Hereford mother and calf enjoy summer days together on pasture.

Healthy milk and beef aside, the purest joy of raising your own cattle is that you do not need to cause suffering in the final days of an animal to feed your family. Pasture-raised beef is equally as tasty as the feedlot variety and, as an added benefit, is one-third to three times leaner than grain-fed beef.

Here's more great news. Beef raised on pasture has two to four times more cancer-fighting omega-3 fatty acids, plus more vitamin E, beta carotene, and folic acids than grain-fed beef. Recent studies also show that CLA (conjugated linoleic acid) and TVA (trans-vaccenic acid) are present in grass-fed beef but may not be found in the grocer's version. These two acids have been flying off the health food store shelves for the last ten years as supplements to fight cancer and cardiovascular disease—two illnesses that have plagued our nation since the industrial era.

Raising cows is to your benefit no matter how you look at it, which breed you choose, or what your preferred purpose might be. You'll find a rich and rewarding experience plus far healthier food for everyone in your family.

Choosing a Breed to Raise

Cattle come in a wide variety of breeds and crossbreeds classified as purebred dairy breeds (registered or non-registered), meat breeds, crossbreeds, dual-purpose breeds, or miniatures.

Dairy and beef breeds are just that. They may have been developed over a few years or a few centuries and they have one specific purpose—either to produce milk at top quantities or to grow quickly with a high meat-to-bone ratio.

Crossbreeds are used most often as beef cattle. These cattle have been bred purposefully—either to thrive in their environment or as an attempt to create a new and better meat breed. Dependent on your objective, a crossbreed might be the best animal for your money. Especially so if you aren't interested in establishing a show-quality or registered herd.

Finally, and worthy of consideration, are the smaller dual-purpose cattle and miniatures. In every other section of this book I've avoided the novelty, the toy, and the fancy breeds, but in cattle the dual-purpose and miniature cattle cannot be ignored based on their service to the small family farmer.

Dairy Breeds

Before you run off to buy a family milker, consider this: the average dairy cow provides twenty to twenty-four quarts of milk every day, three hundred or more days of the year. Consistent yields are realized by keeping one of the top six milk breeds.

Standard dairy cows mature between 900 and 1,500 pounds; bulls and steers, between 1,500 and 2,000 pounds. The most popular dairy breeds are, listed from smallest to largest, Jersey, Guernsey, Ayrshire, Milking Shorthorn, Brown Swiss, and Holstein.

Your decision on which breed to raise might be made based on local availability, quality of pasture, and climate conditions. While all milk breeds are hardy enough to thrive in most North American climates, some may do better than others in your area. Discussions with local ranchers, veterinarians, or your feed supply store are highly beneficial.

The two smallest dairy breeds, the Jersey and the Guernsey, each have an interesting trait that you should know about before selecting them as your breed of choice. Jersey calves sold for beef will net low prices should you plan to sell them. Even though the beef is as tasty as any other, the fat of the meat is yellow, which the marketplace misjudges as "less than fresh." Even a Jersey cross shows yellow-colored fat. The Guernsey, on the other hand, has milk that is a slightly yellow to cream color. This shouldn't present a problem unless you have some real finicky eaters at your dining table.

Beef Breeds

Although any breed of cattle can be raised for beef, some breeds are widely recognized for their good mothering instincts, fast growth, having small calves (small calves equates to easier births), and climatic hardiness.

Hundreds of beef breeds and crossbreeds exist in North America and any purebred or crossbred calf will yield delicious, nutritious, and economical beef under your control within just a few months. If your plan is just to raise one or two calves for the freezer, check the sales of breeds readily available in your area and choose the healthiest calf you can find from the most knowledgeable or recommended seller.

Generally speaking, beef and crossbred calves are forty-five to fifty pounds at birth and kept on pasture with their mothers for the summer and most of the fall. They will have cost little to raise but will provide 320 to 380 pounds of beef for the freezer given that they have reached their seven-month potential of 600 to 650 pounds live weight.

Dual-Purpose and Miniature Breeds

Although there are many dual-purpose breeds in North America, the most popular is the Dexter breed. These smallish cattle reach maturity at 750 and 900 pounds, standing thirty-six and forty-two inches tall at the shoulder, for cows and bulls respectively.

A good Dexter cow will supply one to three gallons of milk per day (a manageable amount compared to her Holstein counterpart of five to six gallons daily). Her offspring at seven to eight months of age will grow to be 350 to 500 pounds, 55 to 60 percent of which will end up in your freezer.

Although the Dexter is a small breed, it isn't considered a miniature by all breeders. In fact, the rules and boundaries between dual-purpose and miniatures are somewhat fuzzy across developers and registries.

Miniature cattle are not the latest fad. Documentation of smaller cattle has been noted since the late 1960s, and at present time there are nearly thirty registered breeds—ranging from the original Dexters and Zebu (from Ireland and Mexico respectively) to the American-made Lowline (a miniature version of the Angus).

These cattle are a viable, useful, and productive alternative for the small families who keep them. It is no longer necessary to be overwhelmed by the massive output of a full-sized breed or to be stuck trying to find friends to share the bounty of milk and split a side of beef with you.

Temperament and Training of Cattle

Dairy cows and beef cattle are herd animals. They settle into a new home most easily when they are with their own kind and have less tendency to be nervous. A lone calf will bond with you and consider you one of the herd if you are quick to understand its needs and its nature.

Calves and mature cattle can be timid if they've been raised on pasture and are not accustomed to human interaction. Even if they were, you as the new owner will not be "their" human until they have spent ample time with you. As long as they have been treated fairly in the past and you give them a little grace and space in the beginning, you'll have them ambling over to greet you in the field or pen in no time.

All cattle have vivid, long-lasting memories that serve them to associate both painful and pleasurable situations with a person, place, or inanimate object. You can tap into that ability once you understand it. You can train youngsters to respect the electric fence or give you space, for instance, as well as teaching them a call word that will have them running back home for grain at top speeds.

The best manner to tame and raise cattle is with a gentle and steady temperament. Cow and steer alike can neither tolerate nor learn to trust the human who is flailing his

arms, screeching and yelling, or rushing at them. You should not be so mild-mannered that you let them walk all over you either. Once you've gained their trust, the next step is to assert and establish yourself as a respect-worthy herd "boss."

Every calf, heifer, steer, and cow also has his or her own self-defined comfort zone. By respecting you, they will understand and learn your own comfort zone and therefore will be less likely to crowd or push you, step on your feet, or trample you in a moment of panic. To teach them about your boundaries you'll have to give them a sharp rap on the nose or the hindquarters with a small stick. This not only gets their attention but also lets them know that they have come too close or that their behavior is unacceptable.

If an intentional, well-placed smack doesn't stop them in their tracks, you haven't hit hard enough. These are heavily muscled animals with thick hides. At 1,000 pounds, they can certainly push you over or crush your entire foot if they are not taught to respect your space. I never condone abuse of an animal, but cattle can easily put you in the hospital, in a cast, or in a wheelchair. One sharp smack does not constitute abuse.

One final point about the temperament of cattle: their shyness can induce panic, and there is nothing safe about a half-ton animal in a panic. Calves panic if they cannot find their mothers and will run circles, risk their own lives, and cry for hours in frustration. Mature cattle will panic if cornered. In that situation and without being controlled they will mow down anything or anyone to get themselves out of the fearful situation.

Train a calf to lead by a halter and you can keep that panic at bay at any age. They will have learned that once in the halter their panic serves no purpose—the human on the other side of the rope is in control. Even with an animal of this size under control, you should never let your guard down. Keep a plan of escape and stay out of the kick zone whenever you're working with one, on or off the halter.

Designing Your Small Farm Strategy

Milk and Meat Yields for Cows

Although widely ranging in size, approximate yields to be expected from raising cattle are as follows:

- Dairy—Average milk supply of 1,500 to 1,800 gallons per year.
- Beef—60 percent of live weight can be expected as dressed and packaged beef.
- Dairy Breeds Raised for Beef—A dairy heifer calf, raised just seven months, weighs between 350 and 450 pounds (netting 190 to 260 pounds of beef). If you raise her another year to 700 to 1,000 pounds, your freezer beef doubles. Add 30 percent if raising a dairy steer for beef.
- Beef Breeds Raised for Beef—A heifer calf at weaning will weigh between 450 and 600 pounds (netting 270 to 360 pounds of beef). Raised to maturity, the heifer weighs 900 to 1,000 pounds. Again, add 30 percent when raising steers.

A few options in raising cattle for a food source are listed below. Choose or alter one of the strategies below to suit your needs, space, and time. Keep in mind that the family milk cow is the animal that, once owned, must be tended to on a schedule. Twice daily milking, twelve hours apart, seven days a week. To put her off is to decrease her value and in some cases jeopardize her health.

- Keep and breed a dairy cow every year for up to six gallons of milk per day, plus one calf to either sell or raise until late fall for approximately 250 pounds of dressed beef in the freezer.
- Keep and breed a dairy cow every year and once she has calved, purchase up to three orphaned calves for her to raise with her own. Your cow will produce enough milk to nurse four calves. Once they are weaned, they can be sold or put to pasture until late fall and you'll still have an ample milk supply for your family for another seven months. These four calves plus six months on pasture could yield 1,000 pounds of beef.
- Keep and breed a dairy cow every year with a beef breed bull. The resulting calf will be smaller at birth but quick to grow on pasture. His crossbred capability could net you 350 pounds or more of beef for the freezer, plus all the milk your family can drink for seven months or more.
- Keep one or more beef cattle on pasture and butcher when the season ends for a quick freezer full of beef (an average of four hundred pounds of dressed beef). Alternatively, winter the animal and raise him on to next fall to double your yield for less than the cost of a hundred square bales of hay.
- Raise dual-purpose or miniature cattle for realistic and manageable output of milk and/or beef, especially if available pasture is minimal.

Tucked away safe and sound, this young calf is living in perfect conditions.

Pens, Pasture, and Shelter

Newborn calves should be kept in a warm and clean barn or shed for at least three weeks. This small calf has the potential for many stress-related sicknesses severe enough to take her life. Ensure that the calf cannot escape, cannot damage herself within her surroundings, has clean bedding in a draft-free enclosure, is treated gently and compassionately, and is receiving adequate nutrition.

The started or newly weaned calf would also have the best start on your farm when kept in a barn for a week or more with similar conditions. Although young dairy calves are at greater risk than beef, both will attempt to escape back to their dams.

Outdoor Pens

Should you choose to move or start an older calf in an outdoor pen, the fencing around the pen will need to be both tight and tall, especially if you are only raising one calf. You'll also need to provide a shelter from cold winds and scorching suns, but the shelter can be moved onto pasture with the calf when you decide to do so.

The standard confinement pen built to take a weaned calf to seven months of age should be at least 1,000 square feet in size. The perimeter should be constructed of thinly spaced, well-supported wire or wood planks. If you plan eventually to pasture the calf, wait until she has settled into her new farm and owners before training her on the electric fence.

Additional Calves

Plan on raising more than one calf in an outdoor pen? You don't need to double the space for every calf you add. One thousand square feet for the first calf plus 250 square feet for every additional calf is all that is required. Be sure to increase bedding and shade area for every calf added as well.

Inside the pen include a three-sided and roofed shed in one corner for bedding, a hay manger, and a salt/mineral block. The shed should be at least one hundred square feet in size and cleaned daily. The water bucket or tub can be set into an adjacent corner and blocked in at eighteen to twenty inches off the ground to ensure the calf doesn't spill, step in, or soil the water.

The Pasture

The pasture requirement for growing cattle is one and a half to two and a half acres each. Miniatures only require one half acre each. In winter months or slow growing seasons you'll need to supplement pasture with hay, and in some situations, a bit of grain.

All cattle require a shaded area. This could be a large stand of trees, but it is better if they have access to a roofed, three-sided shelter. The shaded area needs to be dry at all times and checked regularly for waste removal. Cows, steers, and calves should never be expected to stand or lie down in wet or filthy conditions.

If you have adequate acreage, employ rotational pasturing. Controlling which area of the land your cow pastures will ensure that new growth is eaten evenly.

Pasture fencing will need to be reinforced by electric wire. Cows are quite capable of mowing over any flimsy barrier in a panic. Even a herd that has been taught to respect a powerful electric fence will have heifers in heat eager to crash through in search of a mate. Woven wire cattle fence, forty-seven inches high with a strand above and below of electric fencing, has worked well on our farm.

Feed Requirements

Cattle do well on good pasture with a mineral salt block. Seldom is anything more required. Calves, dairy cows, and cattle en route to market are the exceptions.

Hay that is suitable for cows and cattle is a combination of legume (alfalfa and clover) and grass hay. If the legume quantity in hay is inadequate you can add a little cottonseed, soybean, or linseed meal to their diets until you find a better hay provider. Talk to your local feed supply store or veterinarian for recommendations or advice if you're unsure about the quality of your pasture or hay.

A salt block intended for cattle or a mineral salt block combination ensures your cattle are getting the required amounts of trace elements and vitamins. If your region or pasture is known to be deficient in iodine and selenium it may be added in the local salt blocks you purchase.

Grain and sweet feed (grain mixed with molasses) are used only for weight gain or to supplement a very poor-quality hay. Corn, milo, oats, wheat, and barley are all acceptable grains for cows when used in moderation.

A producing dairy cow is given one half to one pound of 16 percent protein grain for every quart of milk that she produces, after the first week of calving. The quantity of grain required to keep her producing to her maximum will change throughout the year. Keep barn records to monitor and adjust grain ration accordingly.

Hay fed to dairy calves should be fine-stemmed and leafy. Almost any other good quality hay is fine for the rest of the herd. By the time a dairy calf is three months old she should be eating two to four pounds of growing ration containing 15 to 18 percent protein and three pounds of hay per day for every one hundred pounds she weighs.

Beef calves are usually on the field and following their mother's example of grazing. Until they are weaned they are not likely to have ever eaten grain. Grain is usually reserved for beef steers and heifers to help them achieve their top weight before going to market. Changes to their feed should be gradual and never should they be fed a diet consisting only of grain.

Water

All cattle should have access to fresh, clean water at all times. There is never an exception to this rule. One beef breed steer or heifer will drink twenty-five to fifty gallons of water a day; a dairy cow, from fifteen to twenty gallons. This changes with the seasons, the fresh-ness of pasture, and whether or not the cow is in production.

Winter Feeding and Bedding

You'll save money and the last-minute scurry of trying to find hay for sale if you buy freshly baled hay off the field. You can pick it up yourself, one pickup load at a time, or pay someone to collect and deliver it to your farm.

Check hay before storing to ensure it has dried and is not wet or moldy. Moldy hay is of no use to livestock. Hay should be stored off the ground and protected from sun, rain, and snow. Stack it as high as you can on skids and only uncover what you'll need for each day. If the hay is dusty from the field, give it a good fluff and shake before feeding to your cattle.

A mature cow will eat twenty-five to thirty pounds of hay per day during the winter months. The average square bale of hay weighs forty-five pounds. Therefore one dairy cow from November 1 to March 1 will consume 3,600 pounds of hay, or 80 bales (120 days times

30 pounds divided by 45-pound bales). These are assumptive and general guidelines; adjust your winter hay needs to suit your cow and climate.

While you're at the chore of purchasing hay for the winter, you might also prepare for bedding material before the weather turns cold. One cow's bedding matter often accounts for another five pounds of matter per day whether in the form of straw or wood shavings.

Cattle Health

Buying healthy animals, keeping their bedding clean, feeding quality hay, providing clean water, and adhering to a veterinary-approved vaccination schedule will go a long way in keeping your cattle healthy.

Keeping herd, dairy, or barn notes can also be beneficial should any of your animals fall ill. Quick access to vaccination records, changes in feed, and first signs of behavioral changes all help to assess and treat any illness or disease. Also, should you one day plan to sell your cow, calf, or cattle, you'll have records of progress, production, and maintenance readily accessible for the new buyer. (A sample health record for cattle can be found in the forms section at the end of this book.)

Vaccinations

All cattle should be vaccinated by a veterinarian to address nationwide and region-specific diseases. This is not just to protect the animal, but to protect your investment and your own health if you are raising livestock to provide food for the table.

Common vaccinations include infectious bovine rhinotracheitis, bovine virus diarrhea, parainfluenza, blackleg, malignant edema, brucellosis, and leptospirosis. Vaccination schedules may start as early as a few months of age.

Signs of Trouble

If you have spent any time with your cattle you will be the first to notice subtle changes in their health and behavior. Lying down for longer periods, being off feed, kicking at his own belly, lack of interest in surroundings, and restlessness are all subtle signs that the animal is not feeling well.

Cow, Steer, and Calf Vital Signs

- Rectal temperature: 101.5 degrees Fahrenheit
- Pulse rate: Forty to seventy beats per minute
- Breathing rate: Beef, ten to thirty breaths per minute; dairy, eighteen to twenty-eight

- Heifer puberty: Ten to twelve months of age, dependent on breed
- Average birth weight: Thirty-five to forty-five pounds
- Average gestation period: 276 to 294 days
- Heat cycle: Every nineteen to twenty-three days
- Heat period: Twelve to eighteen hours

Breeding for Milk Flow or Beef Calves

Many heifers will reach puberty by their first birthday, but shouldn't be bred until they are fifteen months old. This ensures they don't calve until they are two years old. If breeding a beef heifer, her weight is as important as her age. She should reach 65 percent of the expected mature weight for her breed.

Before breeding, ensure vaccinations are up to date—many of the antibodies in her system will be passed to her calf in the uterus and through first milk consumption. Talk to your veterinarian as soon as you know your animal has been bred to discuss her vaccination schedule, including the newer anti-scour vaccinations.

As long as you aren't planning on raising or selling purebred, registered calves you can breed your heifer or cow with any available bull. The best bull is one that is historically known for throwing small birth-weight calves—especially so if this will be her first calf. The second-best bull is one that lives in the field next door and whose owner doesn't mind a free rendezvous between the two. Taking the time to find an appropriate bull for first breeding is a worthy pursuit. A calf that grows too large in an immature uterus may die at birth or cause physical damage to your heifer.

Signs of Heat

Most heifers and mature cows have no trouble letting you know they are in heat. They will pace at the fence line, bawl to ensure any bull within five miles knows she is ready to breed, and, if other cattle are in the field, either attempt to mount them or allow them to mount her. If your heifer or cow does not display these outward signs, check her regularly for mucus on her back end—a sure sign she is in heat. A heat will only last twelve to eighteen hours. The most opportune time to breed her is in the later half of those hours. Non-bred heifers and cows will return to heat every twenty-one days.

The Pregnancy

Cattle carry their young for nine months (285 days, give or take 9 days on either side). Watch your heifer closely during a first pregnancy for correct weight gain and nutrition. Not only will she be supplying nutrition to the growing fetus, but she'll also still be growing herself. This first pregnancy could affect the remainder of her life and every future pregnancy. Ensuring that she is in top physical health almost guarantees a safe delivery, ample milk for the newborn, and successful breeding in later years.

Nutrition of the Bred Heifer

When pasture growth slows, add a nutrient-rich protein source to her diet. Alfalfa hay is the easiest, most affordable way to add protein, calcium, and Vitamin A to her diet. Throughout the duration of the pregnancy, continue to feed both alfalfa and grass hay.

You only need to supplement her feed with grain if she is losing weight or if the hay you're feeding her is insufficient. Grain will not make up for a shortcoming in nutrients, nor will it keep her warm during cold spells. The digestion of extra hay, not grain, adds warmth to cattle. Your feed store may carry a nutrition-packed feed designed with bred heifers in mind, which may be better than straight grain from the bag.

Awaiting Delivery

Cows that have already calved should be dried off two months before a new calf is due. Heifers might benefit from some practice time on the stanchions. Feed her a little grain, restrain and brush her, then wash her udder and teats. By the time she calves the milking routine will be old hat to her.

While her due date approaches, prepare a place for her to safely calve and gather the supplies you might need on delivery. If this will be your first calving experience, your veterinarian's emergency phone number is a must, as well as a few experienced and local friends' numbers.

Calves can be born on the field or in a barn stall as long as the area is clean and dry. Pasture births should only be allowed if the weather is warm and the cow can be safely alone in a grass-covered, shady spot. Stall births require ample room for the cow to move around comfortably, a non-slippery floor, and clean bedding.

Supplies you might need include:

- Strong iodine solution for the calf's navel
- Clean towels to dry the calf off
- A baby bottle with a lamb nipple in case you need to feed the calf
- Long, disposable gloves in case you need to right the calf in the birth canal
- Fitted halter and rope to restrain or lead your cow or heifer
- Half-inch nylon rope in case you need to pull the calf out of the birth canal

All heifers and cows are different, but any time between a few weeks and a few hours before labor, she may show a full udder with dripping teats, an enlarged vulva, an active tail, and restless behavior. As the moment draws near you may see signs of contractions and her noticeable desire for seclusion. When birth is imminent (twenty minutes to two hours away), a flush of yellow water, the unbroken water sac, or tiny hooves will appear. Most deliveries are trouble-free without human intervention, but if an hour has passed without progress, call the veterinarian for help.

ABOVE: As cows are inquisitive, not much time passes after a new birth before other cows in the herd pay their respects and meet the new addition.

Heifers and cows often lie down for the remainder of the delivery. You can leave her where she lies as long as the calf won't be obstructed during delivery. Once the calf has fully arrived it should either be breathing or your cow should be attentive to the lack of breath. Give her ample time to attend to the task herself, but by all means step in if she doesn't seem to notice the calf. Remove the sac from the calf's face and tickle his nostril to get a sneeze out of him. The sneeze alone should alert your cow enough to take over the remainder of care, but if not, the next step in newborn care is to dry him off completely and gently rub him all over to get his circulation pumping.

If you have time between all this and the moment the calf stands to nurse, wash your cow's udder and teats with warm water only. A calf should be nursing by thirty minutes or, as might be the case with a difficult birth, up to two hours later. You can step in and help cow and calf by standing him up, nose to a teat, and supporting him until he has drunk as much as he will take. If he doesn't appear interested or your cow is being difficult, milk her and feed him by baby bottle.

At some point during the next hour you'll need to treat the umbilical cord and navel stump. The optimum length for a navel stump is three inches, but longer is fine as long as it isn't dragging on the ground. Do not touch the stump, as germs and bacteria are easily passed into the calf's system this way. Dip the stump completely into a small cup filled with iodine. Bull calves will need to have an iodine dip repeated numerous times throughout the first day as they will sully the stump every time they urinate.

Your cow may take a few hours to shed the afterbirth. Once expelled, remove and dispose of it. If it has not expelled completely or at all, do not intervene without veterinary assistance.

First Milk

The first five to six milking sessions after delivery are colostrum, a rich and heavy milk loaded with fats and antibodies intended to increase disease resistance and assist calves with their first bowel movements. One to two good feeds of colostrum will provide all a newborn calf needs to get started on the right foot. By the seventh to eighth day, colostrum is completely replaced by milk suitable for human consumption.

Storing Colostrum

Colostrum can be milked and frozen for many years with minimal loss of nutrients and benefits. Store a gallon or two, clearly marked, and you'll have some on hand to start orphan calves in later years.

If you're raising a dairy cow, the first nurse is your chance to make some dairy management decisions.

Dairy cows generate enough milk to support up to four calves—or at the very least enough for your entire family plus the calf. Consider these options:

- Remove the calf from the cow, milk the cow to feed to the calf by bottle, and commit to bottle-feeding for the next three months.
- Separate cow and calf after first feeding, then allow the calf to nurse from the front teats twice per day while you simultaneously collect milk in a bucket off the rear teats.
- Quickly purchase newborn orphaned calves (from a dairy farm eager to sell freshening calves) and raise them all on your cow's milk by bottle or train the cow to accept each one as her own.

Should you decide to take a more natural route—allowing calf to stay with cow—consider that one of the primary reasons a farmer separates the two is to protect the cow's udder. A cow's udder may be swollen or caked inside after calving. Repeated bunting from a boisterous calf during the first few days of milk flow can cause irreparable damage. Manual milking protects the cow's sensitive udder until her milk flows easily.

If or when you return the calf to cow, begin milking her twice daily, twelve hours apart, to ease udder pressure (one calf will not be able to drink all the milk she's producing) and ensure that every quarter has been emptied.

Dairy Calf Milestones and Management

By one week of age you can add a small piece of starter ration into your calf's mouth after each feeding to help him acquire a taste for it. At three weeks of age he can nibble on fine-leafed hay but he won't eat much of it until he is about eight weeks of age. Sometime between the eighth and twelfth week he can be fully weaned from teat or bottle.

As long as the calf is doing nicely by three weeks of age you can attend to horn buds, scrotal sacs, and extra teats.

Bull calves, once castrated, become steers that are easier to manage and provide tastier beef. Castration is a simple task, causing only minimal discomfort to the calf as long as it is attended to early in life. Using an Elastrator, a tightening ring is attached over the scrotum, which causes the testicles inside to die. No bleeding, just a small tender area that disappears within a month's time. Staff at the feed store—or wherever you purchase your Elastrator—can instruct you on precise use, but instructions also come with the device.

Unless the calf you are raising is a horned-breed purebred, it is best to remove horns when the calf is young. Up to three months of age an electric disbudding iron may be used and takes just a few minutes per calf to perform. Caustic paste can be used on newborn calves (up to three days) but this method isn't without tragedies and is far less popular than the electric disbudding irons of today.

Some heifer calves are born with five or six teats instead of four. The extras serve no use and might even cause her trouble as she ages. Between two and four weeks of age you should be able to easily tell which are the main teats and which are extras. Extra teats are most often found close to one of the main four teats. To remove one, disinfect your hands, scissors, and

Just a little more than eight weeks after calving, this cow is in heat and ready to be bred again. Standard re-breeding practice is your best bet to have a calf on the field every spring.

the teat, then snip from front to back (lengthwise with the body frame) at the point where teat meets udder. Nothing more than a swab of iodine is required after removal. (Do not perform this task if you are unsure or if the heifer calf is any older than four weeks. Call a veterinarian.)

Vaccinations

Within your calf's first month, schedule an appointment with your veterinarian for first vaccinations. Some are given as early as two months of age and may include selenium injections to ensure your calf's nutritional needs are met.

To Breed Again

Your cow will soon come into heat again. You'll know when she's ready to be bred by her usual signs. She may also give a lot less milk on the day she's in heat. Cows are rebred on their first or second heat sixty days past calving. This breeding practice ensures she has a full year between calves and the milk keeps flowing for most of the year.

How to Milk a Cow

A dairy cow, like a milk goat, thrives on routine. They need to see you at the same times every day, preferably twelve hours apart.

The major difference between the cow and doe however, is that a cow can refuse you— even when her milk is overflowing and the udder pressure unbearable. She may have a reason to refuse to let down her milk, but you can work around her reasoning even if she doesn't cooperate. The solution is in finding the trigger that counteracts her stubbornness— a natural hormone, oxytocin.

Milk flow will commence within a minute of an oxytocin release, often stimulated by any act that she associates with giving milk. Triggers might be a feed of grain at the stanchions, being brushed, seeing a milking stool or pail, or an udder wash. In fact the very act of washing her udder and teats with warm water will bring on a flow of milk that even the smartest cow cannot stop. The trouble with this flow is that you only have eight to ten minutes before she's back in control of her own body again.

Cleanliness of your cow and all equipment (including hers) should be your top concern. Sanitize all equipment before and after milking, including udder and teats.

LEFT: A good-natured dairy cow will arrive back to the barn, twice daily, and will stand without stanchions while anyone milks her.

The practice and treatment of obtaining and storing cow milk is similar to that of a dairy goat. The exceptions and differences are provided below.

- Cows can be milked in the field, in stanchions, in the barn, tied or untied, in open weather or under the protection of a milking shed. The preference is both hers and yours, but some cows are more particular about the ceremony than others.
- A cow does not need a gentle bump at the end of milking each teat. You will know when no more milk exists in her udder as her teats will lay flat.
- To dry up a cow—two months before calving—stop milking her altogether. Do not attempt to wean her off the process or slow down production over a period of days. Her body will stop producing milk the day you stop milking her and after a few days of discomfort she'll absorb the milk left in the udder right back into her body.

Longhorn and Grass-fed Beef Cattle

Longhorns

In Texas, there are longhorn cattle dotting the landscape everywhere you look. The local farmers call them YO, which is short for "yard ornaments." And that's why most people keep them. They look nice, don't take any real effort or cost to care for, and you can sell a couple a year at a small profit to keep your agricultural exemption. They are hardy, used to drought conditions, eat almost anything, and very seldom have any birthing problems. Thus they are a perfect animal for the hobby farmer.

BELOW: Chile the steer at Oak Creek Ranch.

What exactly are they used for? Very few people eat them, but their hides and horns are used in clothing, furniture, and decoration. The meat mostly ends up in pet food, so you can keep them around to enjoy a long, restful life (as opposed to other cattle that are slaughtered within a couple of years) before sending them off to the butcher. If you're an animal lover, you probably have dogs and cats that need to eat. Since longhorns aren't mass-produced and are kept mostly as YO, they are probably the most humanely raised meat product around. Ironic, since they are some of the least desirable for human consumption.

Longhorns are also very resistant to disease and require a minimum amount of vaccinations and veterinary care, as long as you're not raising them in a northern climate. Longhorns do best in hot and dry areas of the country. We know farmers in Texas that have only vaccinated their longhorns once, dewormed them only every few years, and never had to call a vet for their YO; they've been raising them healthy and without incident for many years. But it's recommended that you deworm them once a year. The initial vaccinations, which you only need to do once, are quite inexpensive.

Longhorns don't need any shelter really, as long as there are trees for them to lie under. Even if you provide shelter, they probably won't use it. Animals with horns that can span eight feet don't really like to be confined indoors. They can mostly feed themselves on grass, except for the winter months. You'll need to provide them with hay during the period when there's no grass to eat. No one yet has been able to figure out exactly how much cattle need to be fed each day. So the best way to do it is to buy round bales of hay and put them in the special round bale feeders to allow the cows to eat as needed. Water, of course, is just as important, and a large water trough should be available at all times.

Longhorns generally calve without incident. The only exception to this seems to be when a larger breed bull mates with a longhorn. This happened to Tom and Maryneil Dance. A big bull from the next farm over jumped a fence and ended up mating with one of their long-horn heifers. When the time came for her to birth it, the calf was too big to pass. The cow was in obvious distress with the calf only part of the way out. This is a very common situation in most cows, but not Longhorns. After pulling on their own for awhile with no success, the Dances finally tied a rope around the calf's legs that were protruding and hooked the other end up to their all-terrain vehicle. They slowly eased it forward and the calf popped out, no worse for wear. While this story is uncommon for longhorns, it's enough to reinforce our aversion to breeding. Having an off-farm job would make it all the more likely that you wouldn't be around to deal with an emergency like this and it could quickly turn deadly for the cow and calf.

Grass-fed Beef Cattle

If you want to add diversity to your farm operation and want to try to turn a decent profit, you might consider adding a small-scale grass-fed beef operation. It's also a good way to improve your health if you're a meat eater, as grass-fed beef is lower in fat and has more micronutri-ents and B vitamins than the corn-fed beef you can get in the store. But be sure that you have adequate pasture for the cows to be rotated. If you don't have enough pasture for them, you'll just throw all your profit possibilities away on feed and you'll subject your animals to greater risk of health problems.

Stick with buying steers (castrated young males) that are about two years old. Getting into breeding is more trouble than it's worth unless you decide to fully dedicate yourself to it down the road. Buy from a reputable local farmer and start with just two or three steers. Certified cattle are recommended by most farmers, but they are more expensive. Most

people buying meat at the farmer's market don't care if it's certified, only that it's grass-fed and humanely raised. Angus is the best breed for a beginner and provides the meat most people are used to buying. You keep each animal for about two years and then send it off to be processed. So you'll want to buy a couple more each year to keep your yearly supply producing.

Grass-fed beef can be a sustainable part of your farm. But you must manage them carefully. You'll need several pastures that can be closed off. Usually this can be done economically with electric fencing. And you'll need to make sure you can get water to each pasture. You'll want to rotate your cows so that they are never on the same pasture twice in one year. Like everything else we suggest, start small until you know how many cows your pastures can sustainably support. One cow to three acres of pasture is a good start. Wait until they've grazed it thoroughly and move them on to the next pasture. Some people have one cow to two acres, but why go for more until you're sure about your own pasture's health?

Grass-fed beef also goes hand-in-hand with a mobile chicken coop or chicken tractor operation. Rotate the chickens around following the cows. They spread the cow manure and eat the harmful bugs (including fly larvae), all the while producing their own manure as they go. Their scratching of the dirt aerates it and works in the nutrients. If you have enough room, you can also follow the chickens with a field crop, like corn. Let the cows back in after harvest to pick through the leftovers before moving them onto grass again and you've got a healthy rotational system. But planting field crops will require that you go back and seed grass, so it's not a completely self-sustaining part of a rotational system.

Since grass-fed beef is meant for human consumption, there are government regulations that you have to follow. While some beef is labeled organic, there really is no such thing as organic beef. All cows, by law, have to be vaccinated against certain diseases, including black foot. So while your organic steak may not have any antibiotics in it (although there might even be traces of that), your beef has definitely been touched by the pharmaceutical industry. And cattle need to be dewormed. You're basically putting poison in their bodies to kill the worms. While this is allowed for an organic label, it's hardly a natural ingredient.

Cattle must be slaughtered at a government-certified slaughterhouse. And those are getting harder and harder to find. It's unfortunate when you have to truck your cattle a couple of hundred miles to be slaughtered. So while a grass-fed operation decreases the carbon footprint of raising cattle, it doesn't eliminate it. But it's been shown that rotational beef operations can actually create a good carbon balance by returning oxygen to the air through the increase in the health of the grasses. As we've stated before, environmental purity is pretty much impossible on a working farm, but it's worth working in that direction.

Can You Make Money on Grass-fed Beef?

By now you know that there are endless variables in farming, including the market for cattle that year and the availability of hay. There will be lots of up-front costs that should be factored in over many years, like fencing and any infrastructure you lay for water delivery. As long as you stay small, grass-fed beef can be a nice supplement to your farm income. But it's also the perfect example of why you need diversity. As with all the other enterprises on your farm, it's virtually impossible to make a living by raising beef alone without becoming a factory farmer or having many hundreds of acres of healthy pasture. And keeping only cows is a good way to exhaust the natural resources on your farm. Here we provide an approximate breakdown of cost and profit potential for one cow from purchase to sale to consumer on a small-scale grass-fed beef farm:

Costs

Steer—$600 (although they can cost $1,000 or more sometimes depending on the market at the time).

Transportation—If you don't have a trailer, it'll cost at least $50 to have the steer delivered and $50 to have it shipped to the processor; $100 total.

Vaccinations and veterinary—One-time vaccination of about $100, plus once a year deworming of $20; $140 total for two years.

Supplemental hay in winter—$200 per year = $400 total.

Fly control—$25 per year = $50 total.

Processing = $400

Total Costs = $1,690

Return:

One animal will be processed into about 325 pounds of meat (more or less 25 pounds).

Selling direct-to-consumer at farmer's market (average of $6.50 per pound) = $2112.50.

Profit = $422.50 (cut that in half if you are selling it wholesale)

So you can see now why farmers are constantly straining their farm system by adding more and more cattle to make a decent living. Breeding your own cattle raises your profit margin, but creates other problems and costs. One bad drought or one infectious disease and you're wiped out. It's the same unsustainable cycle you see in most farm operations and it's the government that usually steps in to help with farm subsidies. But this just drives prices down and the unsustainable cycle is perpetuated.

Horses

WHAT KID HASN'T GROWN up dreaming of owning a horse? They are the largest creatures that we can still consider as our pets. But anyone who's ever had or worked with horses can attest that they are not pets at all. They are companions and friends. In many cases, they serve as our equals and complement our own abilities. The connection between horse and rider is powerful. So much so that horses are used very successfully in therapy. Horses will work with you. If you have hundreds of acres to tend to, a horse is the very best off-road vehicle you could have. Horses don't get stuck or have flat tires. We may need to help them survive, but only because we've decided to confine them to our farms. Given enough freedom to roam, horses thrive in the wild.

Horses are the one exception to our low-maintenance, low-cost rules of hobby farming. They can be relatively low-maintenance, but most of the heath problems we've over-heard every owner bemoaning are those of their horses. And it's not cheap to feed a horse in winter or to get veterinary care for it. Unlike other animals that produce by-products (like chickens' eggs), horses rarely help offset the cost of their care. But frankly, we owe it to horses. They've been with us every step of the way since we set foot on this land (and even before), carrying our burdens, plowing our fields, and fighting our wars. Without horses, it's hard to imagine that we would have been successful at all in the New World. For that reason alone, it's worth the comparably small effort and money it takes to keep a horse happy and healthy. There are tens of thousands of horses that don't have proper homes or are being crowded out of the last remaining wild places; giving one a home is a noble endeavor.

Horses can live very happily right along with your cattle, donkeys, or llamas. They don't require any extras except perhaps some higher protein feed in the winter. As we mentioned before, they can shelter like the other animals we've discussed in three-sided sheds. But if you live in a very hot or very cold climate, it is best to have a barn with stalls. This gives you a way to get them out of the heat in the summer and possibly run fans on them to keep the flies away. And they can warm up and have easy access to food and water in the winter. They should be dewormed twice a year. It's recommended by most veterinarians and the makers of the medication that you deworm every six weeks. We feel this is excessive, but you'll want to follow your vet's advice as worm problems vary in different climates and parts of the country. You can use the same oral medication you use for the donkeys. And they need mineral and salt licks.

Horses, like donkeys, can founder, which is a painful condition of their hooves that can cause lameness and death. This usually happens after a rainy period that's just followed an extended dry period. The rich, sugary green grass is what makes them founder. This is another good reason to have a barn with stalls as you can control their access to the rich grass. Avoid feeding them too much grain as this can also contribute to hoof problems.

Horse hoof care and shoeing is of utmost importance. Unfortunately, this is not a task that you should do on your own. Call in a farrier. You may have a retired horse that doesn't wear shoes and you may be able to take this task on yourself eventually (it's the same process as that of a donkey), but a farrier will be needed until you are completely comfortable. Farriers are also a valuable source of all information on horses. They spend their time on other people's farms seeing and hearing what works and what doesn't. So they're worth the money in more ways that just putting shoes on your horses' hooves.

Fly Predators

Horses are often plagued by flies in the summer, as are cows. They swarm around their eyes and any open cuts. Traditionally, they've been controlled by using a fly spray you can buy at your local feed store. But these fly sprays are like spraying a pesticide on your animal. And many horses don't like them. Fly predators are a natural form of control. You can order them online. They are small, flying insects that feed on the larvae of flies, thus stopping their reproduction. You spread them around fresh (wet) piles of rotting manure, wet hay, or feed, where flies reproduce. Within thirty days, you should see a big difference in your fly population.

Another worry with horses is that there are many plants, trees, and weeds that are toxic to them. The list is quite extensive. Before getting a horse, you need to have your extension agent come to your farm and help you identify the offenders that need to be taken out. And unfortunately, herbicides are typically the only real option. You can try to mow your grass and replant heavily with healthy grasses. But get rid of them you must. There are many sad stories of horses dropping dead after grazing a small patch of the wrong food.

Unless you've rescued a horse that's injured, riding is good for you and the horse. Not only will this keep you both more in shape and healthier, it's the best way to connect with your horse and spot any problems it might be having. Riding is a social outlet for both rider and horse. Michael's mother Maryneil has a group of women friends that get together regularly to ride. For farmers who sometimes live far away from a city, riding with friends is their main social outlet and the horses generally enjoy being around each other too. Riding is therapeutic and is used more often now in programs that provide disabled children a sense of freedom and empowerment. This freedom is just as liberating for the healthy. Learning to ride is not an adventure you should undertake on your own. Failure to use the proper techniques in saddling and riding a horse can easily lead to serious injury for you

RIGHT: You can tell this horse has foundered by the tell-tale signs of nail holes in the hoof wall where the founder boot was secured.

Maryneil Dance exercises her horse, Rio, in a round pen.

or the animal. Find a stable that offers riding lessons or hire an instructor to come out to your farm and work with you and your horse.

A horse needs to exercise regularly or it risks injury. To do this, many people exercise their horse in a round pen for about thirty minutes, three times a week (although many horses get by with much less). It's also a valuable training tool and a way for you to connect with your horse without riding it. You should read a few books on proper technique and watch a few of the many videos available online. Lunging is exercising a horse in an enclosed area by encouraging it to walk, trot, and canter around you. Basically, you will alternate walking, trotting, and galloping at various speeds and different directions while teaching the horse voice commands. A round pen doesn't need to be fancy. You can just hook panels of fencing together into a circular pattern big enough for the horse to run around at a full gallop. Just make sure there are no sharp points that a horse could be injured on. Excercise the horse for several minutes one way and then switch directions in order to work both sides of the animal's muscles equally. Except for riding a horse, there's nothing more thrilling than standing in the middle of a round pen and watching a horse run full out in a circle around you. It's both frightening and exhilarating; and the horse loves to stretch its muscles in a controlled environment.

After riding or exercising, horses should be cooled down with a long walk. Feel the horses chest, between it's front legs. When it has cooled down, it's safe to hose the horse off (as long as its not freezing outside), wipe it partially dry and turn it out onto the pasture or into its stall.

Don't let the cost and time discourage you from keeping horses. You just need to keep the endeavor in the proper perspective and make sure you truly have the means to take care of a horse. You're giving the horse a good home, which it deserves, and you will connect with it on a deeper level like no other farm animal.

Equipment

Every horse farm needs certain fundamental pieces of equipment. Most of these are self-explanatory, so here I'll just offer a series of lists as inspiration. The three "big ticket" items, tractors, trucks, and trailers, are discussed in their own separate sections below.

Grooming kit: Hard and soft brushes, curry, mane and tail comb, hoof picks, scissors, fly spray, detangling spray, bath sponges, bath bucket, shampoo, sweat scraper, shedding blade, towels, clippers, braiding kit if needed. Ideally, keep a separate kit of basic brushes for each horse to avoid cross-contamination of contagious skin conditions.

Equine first aid kit: Bag Balm, antibiotic wound ointment, icthammol, Epsom salts, VetWrap, roll cotton, non-stick wound pads, standing bandages, poultice, iodine scrub, scissors, Banamine, Bute, latex medical gloves, twitch, thermometer, weight tape, list of each horse's normal vital signs.

Barn cleaning tools: Wheelbarrow, manure forks, broom, rake, muck buckets, scoop shovel, bucket scrubbers.

Tack: Saddles, bridles, girths, martingales, boots, polo wraps, bell boots, a variety of bits, saddle pads, halters and lead ropes, whips or crops, lunge line, lunging cavesson, lunging surcingle and side reins. Choices in tack are highly individual depending on discipline and personal preference, and you may need more or fewer items than those listed here.

Boots and wraps: Bell boots, galloping boots, polo wraps, standing wraps, pillow wraps.

Horse clothing: Turnout blankets, stable blankets, turnout sheets, dress sheets, fly sheets and masks, fleece or wool coolers, quarter sheets, shipping wraps.

English tack.

Feeding equipment: Buckets, grain feeders (stall-mounted or flat pans on the ground), measuring scoops, scale for weighing grain and hay, scissors to cut hay string, hay nets, large water trough, heated buckets, water tank de-icer, rodent-proof containers for grain storage, pallets for short-term hay storage.

Types of Horses and What to Look For

Breeds

There are literally hundreds of horse breeds across the globe. A multitude of books focusing on horse breeds will show you photos and describe a vast array of horses. Here I've selected the most common pleasure horse breeds in North America, the ones most often owned by amateur horse owners.

American Quarter Horses, Appaloosas, and Paints

The American Quarter Horse is the most common horse breed in the United States. Originally developed for working on cattle ranches, these horses are typically quiet, intelligent, and trainable, making them nearly an ideal choice for the novice or amateur rider. Note that there are now many different types of Quarter Horses being bred for many different purposes. For riding, you'll want a performance type rather than a halter type.

ABOVE: The American Quarter Horse is an ideal pleasure mount.

The Quarter Horse traditionally belongs to the Western disciplines—it is the rodeo cowboy's horse, the barrel racer, the gaming horse, the roper, the trail horse, the Western pleasure show horse. However, their sheer numbers, popularity, and versatility mean they are also commonly found in the hunter or dressage arenas. You'll often find Quarter Horse crosses mixed with Arabian, Thoroughbred, and draft bloodlines. *American Quarter Horse Association: www.aqha.org.*

I include Paints and Appaloosas with Quarter Horses because they are all considered stock horse breeds—the traditionally Western breeds—and all have a high percentage of

BELOW: American Paint Horses.

BELOW: The distinctive "blanket" coat pattern of the Appaloosa.

Quarter Horse blood. Paints have similar characteristics to Quarter Horses, but with the distinctive white-painted coat colors. (A "breeding stock" Paint is a horse with Paint blood-lines that does not happen to have the white markings. Note also that *Paint* is a breed, registered with the American Paint Horse Association, while *pinto* is a color designation for a white-splattered coat pattern that can be found in horses of many breeds.) *American Paint Horse Association: www.apha.com.*

Appaloosas are a stock horse breed that features another unique coat color pattern, with smaller spots either throughout the coat or concentrated in a blanket over the rump. Distinctive of the Appaloosas are their striped hooves and mottled noses. Appaloosas are known for being hardy, intelligent, and stubborn. *Appaloosa Horse Club: www.appaloosa.com.*

Arabians

Arabians are extremely popular as pleasure and trail horses, and can be found competing in both Western and English disciplines. They are one of the oldest breeds of horses, and many modern breeds can be traced back to an Arabian heritage, including Thoroughbreds, warmbloods, and Quarter Horses. They are willing, versatile, people-oriented partners, and are generally hardy and easy to maintain. They excel at the disciplines of endurance and competitive trail riding due to their stamina and resilience. They're a smallish breed of horse, commonly ranging from 14 to 15.2 hands, but they tend to be strong enough to carry heavier riders anyway. Many believe Arabians to be the most beautiful of all horse breeds.

ABOVE: Arabian horse. (Note that the T-posts in the background should be capped for safety.)

One caveat is that they may be more spirited, hotter, and spookier than some other breeds, so depending on the temperament of the specific horse, they may be harder for a novice rider to handle and ride. They are also very sensitive and intelligent, so beginners may find themselves outsmarted by their own horse. As always, when horse-shopping, assess each animal as an individual, but keep the breed characteristics in the back of your mind.

Another option is to seek out a half-Arabian. Arabians are often crossed with other breeds in an effort to bring lightness, stamina, and elegance to a heavier breed. Arab-Quarter Horses, Arab-Thoroughbreds (technically called Anglo-Arabs), and Arab-Saddlebreds (called National Show Horses) are common. *Arabian Horse Association: www.arabianhorses.org.*

ABOVE: A stunning Arabian-Paint cross.

Drafts and Draft Crosses

Draft horses, as the name makes clear, comprise the heavy horse breeds originally developed for use on farms and for driving. As such, riding is not their intended use, and many may be too heavy or not well built for riding. However, some draft horses with lighter conformation can make excellent riding horses, and their calm and willing dispositions make them ideal for timid or beginning riders. Some breeds commonly used in the United States for riding as well as driving are Percherons, Belgians, Shires, and Clydesdales.

In addition, draft crosses are becoming extremely popular as pleasure and dressage mounts. These are draft-breed horses crossed with a lighter saddle-horse breed, usually Thoroughbred, but sometimes Quarter Horse, warmblood, Arabian, or another light breed. Belgian-Thoroughbred and Percheron-Thoroughbred are especially nice crosses. In a good cross, you'll get the sturdiness and temperament of the draft horse combined with the lightness of bone and athleticism of the Thoroughbred. *Belgian Draft Horse Corporation of America: www.belgiancorp.com. Percheron Horse Association of America: www.percheronhorse.org. Clydesdale Breeders of the USA: http://clydeusa.com. American Shire Horse Association: www.shirehorse.com.*

LEFT: This Belgian draft horse is not especially well suited to riding. Her heavy build, straight back, and low withers will make saddle fitting difficult, while her steep croup, straight stifle, and rather upright shoulder make her gait short, choppy, and earthbound. Her very short, thick neck and short back will make it hard for her to flex, bend, and accept the bridle.

RIGHT: Although the same breed as the horse in the previous photo, this Belgian mare's conformation makes her much more appropriate for riding. She has a much lighter build, with a graceful neck, sloping shoulder, round hindquarters, and a longer back. Her conformation is not perfect, but she looks like an excellent partner for pleasure riding.

ABOVE: Clydesdale horses in harness.

ABOVE: Peruvian Paso horse performing its lateral gait, the paso llano.

Gaited Breeds

"Gaited" means having an extra natural gait in addition to the usual walk, trot, and canter. This extra gait is usually a smooth, fast, easy-to-ride alternative to the trot. For example, Saddlebreds have the rack and Tennessee Walking Horses, the running walk. Other common gaited breeds are Rocky Mountain Horses, Standardbreds, Missouri Fox Trotters, Paso Finos, and Icelandics. Many of these breeds compete on their own breed show circuits or in saddleseat competition, but they are also good trail mounts. The breeds were specifically developed for trail and pleasure riding, and are gaining in popularity among older adult pleasure riders, since their smooth gaits are much easier to ride for people with arthritis or bad backs. They're generally not the best choice for dressage or hunter/jumper riders, since some may have difficulty trotting and cantering instead of performing their usual gait. *Gaited Horses: www.gaitedhorses.net.*

Grade Horses

A "grade" horse is simply an unregistered horse of unknown breeding—the mutt of the horse world. These horses are widely available, generally inexpensive, and often an excellent choice for

LEFT: The versatile American Saddlebred looks equally nice under English, Western, or Saddleseat tack.

a first-time horse owner. Simply due to the widespread popularity of Quarter Horses in the United States, many grade horses are unregistered Quarter Horses or Quarter Horse-type. Draft crosses and Thoroughbred crosses are also common. One thing to watch out for when considering a grade horse is its conformation. Since these horses are not registered or approved by any breed organization, it's reasonable to conclude that the sire and dam may have been an accidental pairing, or one made with less than rigorous criteria. As such, the parents may be poorly matched, or worse, poorly conformed themselves. Read the section below on conformation, and carefully assess your prospect with long-term soundness in mind.

Morgans

The Morgan is an American breed that originated with one specific stallion—Figure, also called "Justin Morgan's horse"—in Vermont in the 1700s. Morgans are a very typey breed, easily identified by their sturdy appearance, arching necks, distinctive heads, and long manes and tails. They are not especially tall, usually 15 to 16 hands, are usually bay or dark brown, and are most common in New England, their place of origin. Their friendly and sensible dispositions make them excellent trail and pleasure mounts. They are often used for driving as well as riding, and may be seen in lower-level dressage competition as well. *American Morgan Horse Association: www.morganhorse.com.*

BELOW: Morgan horse.

Ponies and Minis

There are a variety of pony breeds of varying size and type. Shetlands, Welsh ponies, and Pony of the Americas tend to be the right size for children. Shetlands are very small, and are known for their devious nature, but many an adult rider has fond memories of the devilish little Shetland pony they had as a child. They come in two main types—classic and modern. Classic Shetlands are shaggy and stocky, while modern Shetlands are refined and slender, appearing more like tiny horses than ponies.

Welsh ponies are, simply put, adorable. They range in size, with four different categories. Welsh section A ponies (Welsh mountain ponies) are the smallest at no larger than 12.2 hands; section B ponies are mid-sized at up to 13.2 hands; section C (Welsh pony of cob type) are also up to 13.2 hands, but are less refined than section B ponies; and section D (Welsh cobs) can be quite large, even horse-sized, with no upper height limit. Section C and D ponies and cobs are often used for driving as well as riding. Many show ponies in all disciplines are Welsh. *Welsh Pony & Cob Society: www.welshpony.org.*

Pony of the Americas (POA) is a unique breed that looks like a small Appaloosa. Originated by the accidental breeding of a Shetland stallion and an Arabian/Appaloosa mare, the POA was then developed into a breed in its own right. Ranging from 11 to 14 hands, they make great kids' ponies, despite the fact that they can be a little stubborn, like their larger spotted cousins. POAs are commonly seen in the show ring in both Western and English

BELOW: Classic Shetland pony.

BELOW: Welsh ponies are a common sight in the hunter show ring.

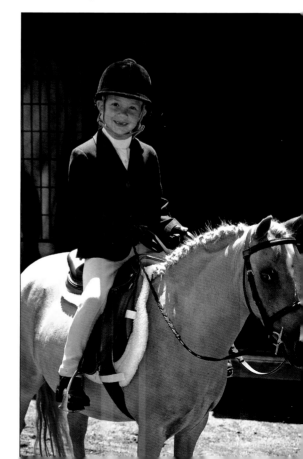

disciplines. *Pony of the Americas Club, Inc.: www. poac.org.*

Pony breeds that can sometimes be larger and therefore suitable for older children or small adults are Connemaras, Haflingers, Quarter Ponies (undersized Quarter Horses), and German riding ponies. The German riding pony is relatively new on the American scene, and is basically a pint-sized warmblood, with the athletic conformation and powerful gaits of a full-sized warmblood in a smaller package. The Haflinger is a draft-type pony resembling a small Belgian horse. They may be used as

ABOVE: Three Haflinger ponies.

driving ponies, but their stellar dispositions make them equally well suited as children's mounts. The Connemara is a large pony breed of Irish origin that excels in eventing and dressage.

Miniature horses are an extremely small breed of horse developed from Falabella and Shetland ponies. Most are too small for riding of any kind, although some are large enough for very young children to ride. Miniatures are best suited for driving or as companions. They are a great choice for an older owner who may find them physically easier to handle and care for, while still reaping the emotional benefits of horse ownership. An owner with only one riding horse may choose to add a miniature horse as a low-cost, low-maintenance companion. *American Miniature Horse Association: www.amha.org.*

Keep in mind that although minis and ponies are smaller and certainly do eat much less than a large horse, their hoof care and veterinary costs will be the same. An additional concern with minis as well as small ponies is their tendency toward obesity and laminitis. A common saying among knowledgeable horsemen is that there are only two types of ponies: those that have foundered, and those that will founder. This may be an exaggeration, but it highlights the fact that smaller equines generally can't safely be kept on grass pastures

full-time, nor allowed unlimited access to hay. If you keep your horses on pasture or feed round bales, think carefully before adding a mini or small pony to your herd. You'll need to consider dry-lotting the

RIGHT: Minis in a driving class.
CREDIT: *SF photo / Shutterstock.com*

pony to limit his access to grass, or using a grazing muzzle so he can't vacuum up the grass as quickly.

Thoroughbreds

Because the racing industry produces around thirty thousand Thoroughbred foals each year, retired off-the-track racehorses are plentiful and easy to come by. Race trainers sell horses who are not working out as racehorses, generally because they're just too slow, but sometimes because they've been injured or are older. It is possible to buy them directly from racetracks or racing trainers if you are confident in your ability to train and handle a young horse. There are several organizations that exist to facilitate this process, including CANTER (www. canterusa.org), New Vocations (www.horseadoption.com), and ReRun (www.rerun.org). If you're not an experienced rider, it's best to look for a horse that has been off the track for several years and has been retrained by a competent trainer.

Many people believe that all Thoroughbreds are "hot" and spirited. This certainly can be the case. However, as with most stereotypes, it's not always true. Many Thoroughbreds turn out to be wonderful, calm, intelligent, people-oriented horses. This is another reason it's best to seek out a Thoroughbred who has already been retrained as a riding horse—you can assess his personality as a pleasure horse much more easily than one that is still racing-fit at the track.

Thoroughbreds make excellent all-around horses for trail and pleasure riding, but they especially excel at the English disciplines—eventing, dressage, and hunter/jumper. *The Jockey Club: www.jockeyclub.com. Thoroughbred Owners and Breeders Association: www. toba.org.*

Warmbloods

Warmblood is not a breed of horse, but rather a type. Warmbloods are sport horses selectively bred in Europe over many generations, to result in the gorgeous, athletic, big-moving dressage horses and jumpers we know today. There are many individual warmblood registries, including Dutch warmbloods, Swedish warmbloods, Holsteiners, Hanoverians, and Trakehners. The term "American warmblood" is sometimes used to refer to generically bred sport-type horses or to draft crosses, but more correctly means a horse registered with the American Warmblood Society or American Warmblood Registry. (The

RIGHT Thoroughbreds are often athletic jumpers.

word "warmblood" originated as a reference to the blend of "hot blooded" horses—Thoroughbreds and Arabians—with "cold blooded" horses—drafts—in their ancestry. Today, though, that meaning has largely been attenuated as warmbloods have become a recognized type in their own right and are no longer literally a cross between hot- and cold-blooded horses.)

A warmblood is an excellent choice for a rider looking to excel at dressage or one of the jumping disciplines. They do tend to be expensive—price should be reflective of quality—so a casual trail or pleasure rider may be able to find a better trained, more appropriate horse at a lower price by choosing a nonwarmblood breed.

Steps in the Horse Buying Process

1. Find a horse—through your trainer, through advertising sources such as print or online classifieds, or by word of mouth.
2. Contact the seller and ask questions. Answer any that the seller has for you.
3. Schedule a visit to try the horse. Bring your trainer. If you like the horse, be prepared to make an offer and put down a deposit.
4. Schedule a prepurchase exam with your veterinarian.
5. If the horse passes the prepurchase exam, bring him home!

How to Find Horses for Sale

- Online sales sites
- Print magazines or local horsey newsletters with classified sections
- Through your trainer—her professional contacts may be an excellent resource.
- Word of mouth—ask around at horse shows and through all your local horse contacts, such as your vet, farrier, and friends. Let everyone know what you're looking for.
- Riding stables—inquire at local riding stables for lesson horses that are no longer able to stand up to the rigors of their professional lives, but might make good prospects for backyard trail horses.
- Sale barns and dealers—tread carefully here, as dealers are in it for the bottom line rather than for the best interests of horses and clients. Bring a trainer or knowledgeable friend, and insist on a prepurchase exam.

Once you've established communication with the seller, have had all your questions answered, and are confident that this horse merits further consideration, set up a visit.

The Role of the Trainer

It's generally best to work with your own trainer or riding instructor when looking for a horse to buy, especially if you're a novice at horse ownership. Your trainer knows your riding ability better than anyone else, and can screen out unsuitable horses. She has a much more experienced eye for horseflesh and soundness (that's why she makes the big bucks, right?) and may have leads on potential horses being sold by other trainers in the area.

Have your trainer help you review any "horse for sale" ads you find, and plan to have your trainer accompany you on any visits. Depending on your own level of competence, you may want to make an initial visit on your own and then bring the trainer along for the second trip if you like the horse. The trainer can ride the horse herself or just watch you ride and offer advice or opinions. Ultimately, your trainer's level of involvement is up to you. Some buyers simply give their trainer a free hand to find, select, and buy a new horse for them. Most buyers, however, want to be the leader in the selection process, bringing the trainer in as a consultant before making a final decision.

Typically, after a successful purchase the buyer compensates the trainer for her assistance with 10 percent of the purchase price of the horse. Be sure you discuss your trainer's expectations for compensation before embarking on your horse search with her.

Trying a Horse

When you go to try out a horse, request that the owner not groom and tack up for you. Doing these tasks yourself will give you a good sense of the horse's overall attitude. Does he flinch or shy when you brush his head? Does he kick out when you touch his flanks or belly? Does he dance around while you position the saddle? Does he pin his ears or threaten to bite when you tighten the girth? All of these are red flags indicating poor training or rough handling, a bad attitude, lack of respect, or in some cases a potential physical problem. You want a horse that stands quietly and patiently while being groomed and tacked, that lets you touch any part of his body and picks up his hooves when asked, and that accepts the saddle and bridle willingly and without a fuss.

Before riding, lead the horse around for a bit and assess his attitude toward you. Is he attentive and ready for your cues? Does he lead well and stand at the mounting block? Will he back up, turn in both directions, and trot in hand with subtle cues?

If the horse knows how to longe, ask the seller to demonstrate this for you. Note the horse's attitude and attentiveness to the handler. Does he respond promptly to verbal cues to

walk, trot, canter, and whoa? Does he show any fear of the whip or aggressiveness toward the handler? Now is also a good time to assess the horse's gaits, with an eye toward the requirements of your chosen discipline.

Never be the first one to ride a potential horse. Always ask the owner or the owner's trainer to ride first. (If the seller refuses to get on the horse—this is a major red flag!) If all goes well, try him yourself. Always ride in an arena or enclosed area when you're on an unfamiliar horse. Note whether the horse stands calmly and patiently at the mounting block, or if he needs to be held. Walk, trot, and canter or lope in both directions. Try several changes of direction at the walk and trot. Does the horse move forward willingly and slow down easily when asked? Does he respond well to leg and seat aids? Are his gaits comfortable and balanced?

If you're considering this horse for a specific discipline, be sure to put him to the test. Try cantering a jump, or even a small course, if you'll be jumping the horse at home. For a dressage prospect, note whether the horse willingly goes on the bit and bends his body in response to light leg and seat aids. For a speed or barrel horse, can he quickly accelerate while staying in control? If you're planning to trail ride, ask if there's somewhere you can take the

horse for a little test ride down the trail. The seller may wish to accompany you on another horse.

One of the most important considerations for a first-time or novice owner is the horse's temperament and behavior under saddle. He should remain calm, willing, and responsive. He should be tolerant of rider errors, rather than becoming anxious or irritable. Pinning ears and a swishing tail are signs of resistance that could indicate either an attitude problem or a physical problem, such as back pain. Obviously, such forms of resistance as bucking, kicking out, rearing, or balking are to be avoided. In theory, a good trainer can help you work through some of these problems—but wouldn't it be better to buy a horse without training problems in the first place? One vice that is an absolute deal-breaker is rearing. Not the little bunny-hop type of rearing, but full-blown vertical rearing. This is extremely dangerous, as the horse can easily flip over backward, either by accident or deliberately, potentially resulting in severe injury or even death to the rider or himself. I would also consider rank bucking to be a deal-breaker. Again, a little resistance in the form of a crow hop is not a death sentence for an otherwise nice horse, but if you find a horse that goes into rodeo mode and is deliberately trying to ditch the rider, pass.

Making an Offer

After riding the horse, you may decide to go try a few others before making a final decision. You can also ask the seller if you can schedule a second visit to ride the horse again. Realize that the seller has no obligation to hold the horse for you at this point, so he may be sold to someone else if you wait and come back later. Nevertheless, buying a horse is a major life decision, so you should not let anyone pressure you into making a choice before you're 100 percent sure.

When you are confident that this is the horse for you, make an offer to the seller, contingent on a prepurchase exam. The seller may accept, reject, or counter-offer. (Note that these proceedings may be handled by your trainer acting as your agent, if you've authorized her to do so.) Once the seller has accepted your offer, you may be asked to put down a deposit. This is a good-faith payment to the seller so she can take the horse off the market with confidence that you will not disappear into the night, and you can be confident that the seller will hold the horse for you until you can schedule a prepurchase exam.

In some cases, the seller may offer a short-term trial period—from a week to a month or so—for you to take the horse home, ride him, and get to know him before making a final decision. If you decide to go this route, be sure to get everything in writing so there are no misunderstandings. The horse should be insured (by you) so that if any accident befalls him, the seller is covered. The prepurchase exam can take place during the trial period as well.

What All Horses Need

Food and Water

Of course, the two basic needs of any living thing are water and food. The water part is simple. Provide your horses with access to clean, fresh water at all times. Period. Scrub and refill stall water buckets daily, and check them often throughout the day to make sure they're full. If a stalled horse empties his bucket overnight, provide him with two buckets. For pastured horses, check the trough several times a day and scrub it each time it needs filling (possibly only every two or three days, depending on the size of the trough and number of horses using it). A splash of bleach (followed by a clean-water rinse) will help prevent the growth of algae between scrubbings. In winter, use heated buckets or trough de-icers to keep the water not only ice-free, but pleasantly drinkable. Horses don't like to drink ice-cold water, so they may not drink enough if it's not warmed slightly. Most such heaters are equipped with thermometers that turn them on in freezing temperatures and off when it warms up, so they're simple and labor-free to use.

Food for horses is a much more complex issue, but the basics of the issues are as follows:

- Hay or pasture should be the basis of every horse's diet.
- Assess the need for grain or other additives on an individual basis for each horse.
- Always feed the best quality hay and grain. Cheap, poor-quality feed is unhealthy and will cost you more in the long run by having to feed more of it.
- Hay or pasture alone may not provide sufficient nutrients. Add vitamin/mineral supplements as needed.

Every horse's nutritional needs are different, so I can't give you a specific diet to follow for all of your horses. Read any of the excellent books available on the subject of equine nutrition, do research online and in magazines, and arm yourself with as much knowledge as possible. This way you can tailor your feeding program to each horse's unique needs.

RIGHT: A trough de-icer keeps water drinkable in winter.

Horse Chores

Daily

Feed (at least twice a day)

Visual inspection of each horse

Administer medications as needed

Clean stalls, run-in sheds, and paddocks

Sweep aisle and tidy barn

Check water and empty, scrub, refill as needed; drain and roll up hose

Turn in/turn out

Groom and pick hooves as needed

Blanket changes

Fly spray, take off/put on fly masks and sheets

Weekly

Rotate pastures

Bleach and scrub water troughs

Re-bed stalls as needed

Clean tack

Feed store run for grain, shavings, etc.

Put out new round bale (if using)

Monthly or bimonthly

Mow pastures

Clean out paddock with tractor

Check and repair fencing as needed

Weed-whack fence lines

Deworm horses as needed

Farrier or trimmer appointments

Buy and put up hay (if storage space is limited)

Purchase supplements

Bathe horses as needed

Check and replace salt/mineral blocks

Turn composting manure piles

Annual or semiannual

Fertilize and re-seed pastures

Add and grade footing in paddocks and run-in sheds

Spread manure on pastures

Build new fences as needed

Build new stalls or shelters as needed

Buy and put up hay (if you have enough storage space for the year)

Install/remove heated water buckets and trough de-icers

Dental appointments

Vet appointments for vaccinations and wellness exams

Companionship

The rare horse can exist happily in a vacuum of companionship. There are horses out there who live solitary lives, in a pasture alone, befriended only by their owners, and they may seem to do just fine. However, the vast majority of horses are extremely unhappy when living alone. Horses evolved as herd animals, and are most content and fulfilled when living in a herd. A solitary horse is ever on alert against danger, never able to let a herd-mate take over guard duty so he can relax fully. Horses in a group spend much of their day

ABOVE: Horses are most content when they have companions.

ABOVE: Companion animals don't technically have to be equines . . .
BELOW: Horses need the freedom to kick up their heels.

socializing, grooming each other, and having discussions about the pecking order; a lone horse is denied these social outlets. Thus, a solitary horse may become stressed, anxious, and bored.

It's ideal to keep at least two horses (although most horses are even happier in a larger group). If you plan to keep only one riding horse, consider finding a companion horse to keep him company. Older horses or those that are pasture-sound but unsuitable for riding are often free or cheap to acquire. To keep costs down, look for an easy keeper, even a pony or a mini (although you may have to closely monitor their grass intake to prevent obesity or laminitis). A donkey can also be an excellent companion with low maintenance costs. They are quite noisy and serve as an excellent guard animal to protect any livestock or poultry you may have. If you prefer not to have another equine, a few goats can serve as an admirable substitute. Goats come with their own set of management requirements and problems, though, so to me it seems best to stick to one species.

Turnout and Exercise

All horses need freedom to wander and stretch their legs and minds daily. Some types of show horses are kept in stalls at all times except during training, but to me this is not an ideal way of keeping horses. It's neither good for their minds and bodies nor efficient for the home horsekeeper. In my opinion, the best way to keep horses on a small scale is 24/7 turnout. It's cost- and labor-effective, and it keeps the horses happy and healthy. Even if you decide to stall your horses for part of the day, be sure to turn them out for at least twelve hours every day.

Shelter

Horses that live outdoors need access to a sturdy, wind-resistant, three-sided roofed shelter to seek protection from wind, rain, snow, sun, and insects. Most states have requirements for this type of shelter. If you have a barn with a stall for each horse, and you're vigilant about bringing the horses inside in adverse conditions, you can get away without having a run-in shed, but it's really easiest to have one that the horses can choose to use at will.

Grooming

Okay, so horses don't need grooming if they aren't being ridden; they certainly won't die without it. But grooming is an important part of keeping your horses happy, healthy, and beautiful. Regular grooming sessions afford you the opportunity to carefully check over the horse for any injuries, cuts, swellings, skin conditions, hoof problems, and attitude changes that can signal a health problem. They are also a chance to bond with your horse, get to know him better, and maintain his training for routine handling. Grooming is a source of joy for most horse lovers and horses alike, which is reason enough to do it. Each time you

My horses have access to their shelter at all times.

ride, drive, or work your horse, you do need to groom him before and after to make sure his skin and coat are clean under his tack and to make sure there are no foreign objects in his hooves.

Training

Yes, all horses need training. Even horses that are not being ridden need training on their ground manners and handling. Each time you handle your horse, you are training him, whether consciously or not. If you're grooming your horse and he pins his ears and threatens to kick you, and you back off, what have you just taught him? You've taught him that he doesn't have to respect you (or anyone else). Next time that threat could become a real kick. The same goes for any situation in which you're handling your horse. He needs to be polite, respectful, and patient when being turned in or out, waiting for his dinner, loading on a trailer, or having farrier or vet work done. These are all training opportunities.

Costs of Horsekeeping

My claim is that keeping a horse can cost less than $100 per month (depending on costs of feed and bedding in your area of the country), and here I intend to prove it. Below is a chart listing all of the monthly and annual expenses for maintaining a horse, broken down by month. Daily expenses such as feed have been calculated by multiplying the amount of feed per day by 30 days (for example, 6 pounds of grain per day, times 30 days per month, divided by 50 pounds per bag of grain, times $18 per bag of grain, equals $64.8, or approximately $65, per month). Annual and bimonthly expenses have been calculated by determining the total per year and dividing by 12 to get a monthly average (for example, 6 shoeings per year multiplied by $100 per shoeing, divided by 12 months per year, equals $50 per month).

The Maximum column lists costs for the most expensive type of horse. It assumes a horse being fed one bale of hay (at $4.25 per bale), 0.5 pounds of vitamin/mineral supplement (at $20 per 50-pound bag), and 6 pounds of grain (at $18 for a 50-pound bag) per day; living part-time in a stall that needs 3 bags of shavings per week at $5 per bag; on a rotation deworming schedule (an average of $7 every 8 weeks); being given supplements totaling about $50 per month; with shoes ($100 every 8 weeks); dental care every six months ($60 every six months); and an average of $200 in annual routine vet care (vaccinations and exam).

The Minimum column lists costs for the least expensive type of horse. It assumes a horse being fed no hay (i.e., pasture is sufficient; of course, in winter you'll have to feed hay no matter what, so this number is really only valid in summer), 0.5 pounds of vitamin/mineral supplement (at $20 per 50-pound bag), and no grain; living outdoors with shelter (no bedding); on a targeted deworming schedule ($20 for annual fecal exam plus 2 dewormings at an average of $7 each); no supplements; with barefoot trims ($50 every 6 weeks); and annual dental care ($60 every 12 months). So following my suggestions for horse care (depending, of course, on your local costs for hay, grain, bedding, and services, which can vary greatly), a single horse can cost you as much as $394 per month or as little as $98 per month, including all annual expenses. If you only consider daily expenses, that "Minimum" type horse only costs $6 per month!

MONTHLY EXPENSES FOR ONE HORSE

	Maximum	Minimum
Hay	$150	$0
Vit/min supp	$6	$6
Grain	$65	$0
Bedding	$60	$0
Deworm	$45.50	$34
Supplements	$50	$0
Hoof care	$50	$36
Dental care	$10	$5
Routine vet	$17	$17
Total	$393.50	$98

All horses should be trained in basic handling, including being caught, haltered, and led easily.

Any rideable horse should have a base level of under-saddle training as well, for his own protection. What I mean is that if something were to happen to you that forced you to sell off your herd, any horse that's not trained for riding has a much greater chance of ending up somewhere unpleasant, such as in a low-end auction ring or on a truck headed for a slaughterhouse in Canada or Mexico. This may sound dramatic or extreme, but it's true. These days a horse with no training has very little chance of finding a loving home. Have your horses trained, at least to walk, trot, and canter under saddle, to give them the best chance of a happy future.

Common Myths About Horse Care

There are many myths that are propagated among the horse community, whether consciously or simply by virtue of the fact that most boarding stables seem to abide by these unspoken "rules" of horse care. In many cases these methods and policies exist for the convenience of the boarding stable's management. That's not to say that a boarding stable's methods are wrong or bad for the horses. On the contrary, there are many "right" ways to keep horses. However, when you move your horses home, you will find that the economics of keeping two or three horses at home are very different from keeping twenty or thirty horses in a boarding stable environment.

For example, in order to preserve their pastures and allow all horses to have daily access to grass, a large boarding stable may need to keep horses in their stalls for most of the day, only allowing them into the pasture for a couple of hours. Letting all the horses roam free would quickly overwhelm and destroy the pastures. The result is that owners who have only ever boarded their horses may believe, because they've never seen another way, that horses need to live in stalls. However, at home, you may be able to let your horses have 24/7 access to the pastures during seasons when

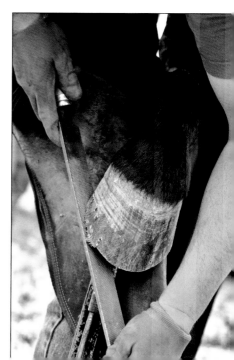

A farrier rasps a horse's hoof to shape it before shoeing.

the grass is growing. There are many such misunderstandings about horse care. The good news is that the truth often saves you money. Read on.

All Horses Need Grain

The truth is that horses evolved to survive on grass and other wild plants. Domesticated horses often lack access to free-choice pasture or hay, and therefore their calorie and nutritional needs must be supplemented with a concentrated, man-made source—grain. If horses have access to free-choice hay or pasture, the majority will keep their weight well and will not need any grain at all. In fact, feeding large quantities of grain can be unhealthy for horses, contributing to gastric ulcers, colic, laminitis and founder, insulin resistance, and many other common equine maladies.

Because hay is expensive and good-quality pastures of adequate size are hard to come by in most areas, boarding stables usually compensate for the lack of forage by feeding grain. When you keep your horses at home, you have the luxury of feeding them as much top-quality hay as they need to maintain their weight. Every horse has a unique metabolism. Some are hard keepers, and will need free-choice hay. Others are "air ferns" and need only a few flakes of hay per day. Watch your horses carefully, and stay alert to signs that they need less or more hay.

It is important to seek out the best quality hay that you can find. The best hay will have a higher percentage of the nutrients that horses need, and will also contain more protein and calories. It is actually a false economy to buy cheap, low-quality hay, since you will inevitably have to feed more of it. Last year, due to the six weeks of rain during hay harvesting season, it was nearly impossible to find good hay. I had to make do with what was available, and the result was that my horses each went through a bale of hay a day in the winter. This year, I've located a source of excellent hay, and I've found that I have to cut back significantly to prevent them from getting fat. Although the price per bale is higher, I am actually saving money by being able to feed less hay.

There are, of course, exceptions. Some horses may need more nutrients and calories than hay alone can provide. Horses that may need grain include: nursing or pregnant mares; growing youngsters; horses in hard work; older horses whose worn-down teeth may not allow them to chew hay efficiently; hard keepers whose metabolism is naturally higher; horses that are ill or recovering from an illness; horses in a rescue situation whose weight needs to be brought up from well below normal. If you do find that you need to provide grain to your horse, be sure to choose a good-quality product. Talk to feed reps, talk to your vet, and talk to other owners in the area. Do not buy the cheapest grain you can find. Like hay, buying cheap grain is a false economy. You will have to feed more to achieve the desired results, and you will be filling your horse with unhealthy sugars and fillers.

In some cases, your grass or hay may be lower quality and may need to be supplemented with a vitamin/mineral supplement or ration balancer that provides the missing nutrients.

Clear signs that your forage may lack nutrients are dull coats, brittle hooves, and weight loss. There are several products available that meet this need. Ask your feed store to recommend a good vitamin/mineral mix to balance out your hay's nutritional profile. I feed my horses Blue Seal's Min-a-Vite Lite. This is a pelleted grain-like product that contains concentrated amounts of necessary nutrients without the bulk and high calories of grain. I feed two cups a day and don't have to worry that my horses aren't getting what they need. In addition, the soil in some geographic locations is deficient in selenium, a necessary micronutrient. If you live in such a region, you'll need to add a vitamin E and selenium supplement to all of your horses.

Horses Have to Live in Stalls

No, they do not. Most horses, in fact, are much happier living outdoors. They can quickly become bored and pent-up when they are kept in stalls for much of the day. Think about it—even if you turn your horse out all day, he is still in his stall at night for up to twelve hours. Twelve hours locked in a box! Many boarding stables offer even less turnout, with horses staying in their stalls for up to twenty-three hours of every day. Psychologically and physiologically, this is unhealthy for a horse. Horses evolved to move constantly, roaming throughout the day in search of forage. They can quickly become anxious and stressed when confined. They are herd animals who are only satisfied when they can interact naturally with a group of other members of their own kind. Their lymphatic systems require the concussion of movement to function properly; this is why some horses become "stocked up" when in a stall for a period of time—their lymphatic fluids literally pool in the lower legs, causing swelling. A horse that is eating hay in a stall stands still for hours at a time. A horse grazing in a pasture walks almost constantly in search of that next blade of grass. This continuous, low-level movement helps maintain muscle tone, encourages circulation to the lower limbs and hooves, burns off excess energy, and improves digestive function. In addition, the pastured horse has the opportunity to gallop, buck, frolic, and play whenever the urge strikes.

You may find that many behavioral problems simply evaporate when your horse moves to an outdoor living situation. Cribbing, weaving, pacing, and stall-walking are all anxiety-caused vices that can be virtually cured by 24/7 turnout. Bucking, jigging, spooking, bolting, and other under-saddle behavior issues are often caused or exacerbated by excess energy. More turnout allows the horse to exercise at will, so he will be calmer and less likely to act up while being ridden. The same goes for horses that are difficult and reactive while being handled.

The ideal situation for most horses is to live outdoors in a group in a large paddock or pasture with access to a shelter, such as a three-sided run-in shed. The good news for you is that it is much less time-consuming and expensive to maintain horses in this kind of set-up. Stalled horses need to be turned out and brought in daily. Their stalls need to be thoroughly cleaned and re-bedded at least once a day, if not more. They will go through three or four bags of shavings a week, at $5 a pop. Water buckets need to be dumped, scrubbed, and filled each day, and probably filled at least two or three more times throughout the day. In contrast, horses who live outside are much more self-sufficient. Because they can come and go from their shelter, it will stay much cleaner. Some owners don't even use bedding in the shelter. I do bed mine with shavings, because I like to know that there's a clean, dry, soft place for the horses to stand or lie down if they so choose. Even so, I have only one shed to clean each day instead of four stalls. I go through one or two bags of shavings per week for all four horses, instead of the eight to sixteen bags I would use if they were all stalled. That alone is a huge savings. I have a large, seventy-gallon trough that I dump, scrub, and refill approximately every three days, saving me lots of time in daily chores.

Another factor to consider is the fact that occasionally, despite your best intentions, you may not be able to get to your chores in a timely manner. On New Year's Day, you may want to sleep in an extra hour or two. Or perhaps there is an emergency that needs to be handled at work and you can't get home until 9 p.m. Any number of unexpected circumstances can arise, preventing you from getting home. Your mind will be at ease knowing your horses are outside with access to grass and plenty of water in their trough, rather than trapped in filthy stalls for hours with no hay and no water.

Now, there are many times when a stall can be a good and useful tool in your horse-keeping toolbox. Although my horses live outdoors with a shed, I do have stalls and I bring the horses in for many reasons. A horse may need stall rest because of an injury. We like to have them all in and handy for a vet or farrier's visit. There may be one horse whose feeding needs are very different, so he has to be separated at meal times. We use the stalls for grooming, tacking up, and untacking. We put the retired horse in her stall while we ride the other two so that she can't run around and injure herself if she gets upset about being alone. We bring all the horses in when the weather is extremely bad—either unbearably hot and buggy in the summer or unbearably cold, snowy, and windy in the winter.

Many home-based horse owners never bring their horses into stalls due to weather at all. They stay out, with shelter available, no matter the weather. Personally, during those frigid January nor'easters when the temperature drops below zero, the snow is flying, and the wind is whipping past my bedroom window like a banshee, I sleep much easier knowing my horses are tucked into their stalls, safe and comfortable. Sure, they won't die being left out. But is that really the minimum standard of care you want to provide for your animals—that they won't die? Better to know that they are not suffering from the cold. The ideal scenario is to have a stall available for every horse, even if you don't use the stalls every day.

All Horses Need to Wear Shoes

There are many good reasons for horses to wear shoes. The main ones are that they protect the hoof walls from chipping and cracking, and they protect the soles from bruising and soreness. What some owners don't realize, however, is that a healthy hoof with a good barefoot trim can serve these functions on its own. Provided the horse has the underlying genetics and good nutrition, and that he is being maintained on a regular basis by a trimmer skilled in barefoot trimming, he can grow a hard, solid hoof wall that will resist chips and cracks. His sole will become callused and thick, protecting the sensitive inner parts of the hoof from contact with the ground. For horses that are in transition from being shod to barefoot, or for horses that are ridden over rocky or hard surfaces, hoof boots can offer additional protection.

Shoes themselves can also be problematic. A horse that pulls off a shoe can do considerable damage to the hoof wall. A shoe holds in snow and ice in winter, leading to a buildup on the sole of the hoof until the horse is literally walking on balls of ice, a dangerous and unpleasant condition for the horse and a difficult task for the human who must chip it out. Similarly, a shoe can hold in dirt, manure, and mud, allowing it to pack deeply into the clefts of the frog and contributing to hoof infections like thrush. An unshod hoof allows dirt to fall out more easily as the horse moves. Some barefoot enthusiasts argue that a shoe prevents the horse's hoof from performing one of its most important functions, shock absorption, by eliminating its ability to expand and contract naturally. In a group turnout situation, shoes can be dangerous; a kick from a shod hoof does much more damage than an unshod one.

More good news for the owner of a barefoot horse: It's less expensive! A barefoot horse may need to be trimmed slightly more frequently (every four to six weeks) than a shod horse (every six to eight weeks), but each trim costs significantly less than a trim and shoeing for a shod horse.

Now, this is not to say that all shoes are bad and that all horses are better off without them. Some horses simply do not have the genetic makeup to grow strong hoof walls. Some have flat or thin soles, making them vulnerable to hoof soreness or bruising. Every horse is unique, and should be evaluated on an individual basis. Talk to your vet and farrier about whether your horse may be able to live and work comfortably and safely without shoes. Horses who are not being worked generally do not need shoes at all. The ideal candidate to go barefoot is a horse who has naturally strong walls and thick soles, and is worked on surfaces that approximate the surface he lives on. That is, a horse cannot be expected to go from a soft, grassy pasture to a rocky, root-covered trail without some discomfort. His hooves simply are not acclimated to it. But a horse who lives in a gravel-covered paddock and who works in a sand arena should be just fine.

Horses Must Be Fed on a Strict Schedule

Many people believe that horses need to be fed at the exact same time every day, or else they will colic, become stressed, or suffer some other terrible fate. It does seem to be the case that horses can "tell time," and if they are accustomed to being fed at, say, 5:30 every evening, they will commence hysterics at promptly 5:31 if they have not been fed. The remedy for this is not, as many assume, to feed on a strict schedule to avoid equine stress, but rather the opposite. If you feed your horses according to a very loose time clock—say, breakfast at any time between 6 a.m. and 9 a.m. and dinner at any time between 4 p.m. and 7 p.m.—they will not come to anticipate their precise dinnertime, so anxiety and stress are vastly reduced if you're "late" with their meal one day. This works especially well if you're only doling out grain at those times, and they have free access to hay or pasture, so they're never actually hungry. Not maintaining a strict schedule has benefits to your personal life as well as reducing equine stress around feeding times. You can sleep in a little on Sunday, for example, or run a few errands after work, knowing that the horses are fine waiting a bit for their meals, and are not pacing the fence line in a frantic state wondering when you will appear.

Horses Have to Be Dewormed Every Six Weeks

For decades, the recommended practice was to deworm all horses every six weeks, rotating through a variety of dewormers designed to kill a broad spectrum of internal parasites. Current research is changing that trend. Many vets now recommend an annual fecal egg count to determine which specific parasites are infecting each horse, and deworming specifically to combat those worms. The concern is that rotation deworming may be creating a new generation of evolved superparasites that are immune to the standard dewormers, much like how overuse of antibiotics in human medicine has led to the creation of drug-resistant superbacteria.

Dr. David Jefferson of Maine Equine Associates explains the thinking behind this new style of deworming:

> Most horses are able to co-exist with parasites. It has proven to be unrealistic to make our animals parasite-free. So all horses carry some worms, but seem to have their own resistance to them. . . . However, there are some horses that don't. Parasites in those horses thrive. . . . The horses that carry this big burden of parasites are called "shedders," and are largely responsible for contaminating the ground with worm eggs. . . . It is the shedder horses that revised worming programs are targeting.

> You can't tell by looking which of your horses is a shedder. It could be the healthiest looking horse in the herd. Fecal exams have to be done on all of your horses to determine who are the shedders. The exams are performed either by veterinarians in their office or they might send the sample to a commercial lab. Shedders are identified by a very large number of eggs per gram of manure. Once identified, the shedders are the ones who get the intensive deworming attention, while others in the barn may need only once or twice a year worming. This system is called targeted worming. It is a program best guided by your veterinarian.

> The object of targeted worming is to cut way back on the amount of worm medicine being used and so lessen the ability of parasites to develop immunity to it. . . . At the moment there are no new wormers being developed, so resistance will become more and more of a problem. (*The Horse's Maine*, May 2011)

Yet again, making this horsekeeping change can end up saving the horse owner money in the long run. While the cost of having your vet perform an annual fecal count on each horse may seem like a high one-time expense, you will save the cost of buying dewormer every six weeks. Discuss this option with your vet to determine if targeted deworming is the right choice for your horses.

Horses on Pasture Don't Need Anything Else to Eat

Some easy keepers may be just fine on an all-grass diet, provided the pasture is well maintained and rotated, with a healthy mixture of horse-appropriate grasses. Unfortunately, many pastures are not equipped to meet the nutritional needs of horses, or they may be overgrazed. Tradition states that one horse needs one acre of grass, but that's only valid in an environment where growing conditions are good. In an arid environment where the grass is poorer and grows more slowly, each horse may need several acres to browse. Most small-scale horse owners simply don't have access to enough good-quality pasture. To offset any nutritional deficiencies in your grass, add a vitamin/mineral supplement to each horse's daily ration. If a particular horse can't maintain a healthy weight on pasture alone, consider adding a high-quality grain or a fibrous nutrient source such as beet pulp, alfalfa cubes or pellets, or simply extra hay fed in addition to pasture grazing.

Horses Are Extremely Expensive

Well, okay, this one is true. Horses certainly are a luxury pet that most Americans simply can't afford. But there is a wide variety of horsekeeping styles that can make horses relatively affordable or place them within the realm of the super-rich. You can keep your six-figure warmblood in full training at a high-end dressage or hunter facility in the suburbs at a cost of several thousand dollars per month. Or you can keep your $500 off-the-track Thoroughbred at home in a rural area for less than a hundred dollars a month. It's true! Less than a hundred dollars a month! It will take planning, know-how, and some sacrifices (you didn't really need that trendy new pair of breeches anyway, right?), but it is possible.

Consider the fact that most middle-class Americans have some sort of expensive recreational hobby. They may be into motorcycles, golf, biking, skiing, snowmobiling, classic cars or racecars, or any number of other pursuits. Owning horses is like many of these hobbies, with one important caveat: The horse is a living being. If a golfing enthusiast loses his job, he can simply cancel his club membership and stop paying greens fees until he gets back on his feet. A horse owner can't simply stop feeding her horse. Yes, in a worst-case scenario the horses can be sold, but in the current economic climate, it is hard to sell even a high-quality horse. One that is old, unsound, or poorly trained may be impossible even to give away. So keep this point in mind as you decide to delve into the world of horse ownership. It doesn't have to be crazy expensive—and I'll give you pointers along the way on how to save money— but it is a commitment. It's not just a hobby; it's a lifestyle.

Grooming and Daily Care

Pick your horse's hooves daily to make sure they stay clean and free of thrush, as well as to check for foreign objects. Abnormal heat in the hooves can indicate trouble, such as an abscess or the start of laminitis.

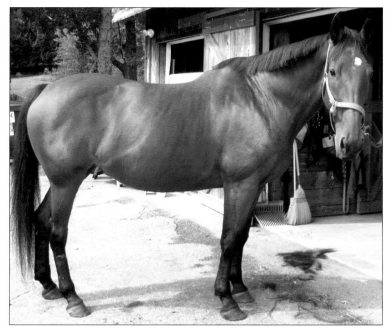

Grooming can be a meditative process, allowing you to "get in the zone" and focus fully on what you are doing in the present moment—forget about bills that need paying and chores that need doing; let go of work-related anxiety and family-related stresses. Just groom your horse. Feel his muscles under the curry. Notice when he responds with pleasure as you find those itchy spots, stretching out his neck and curling his lip. Watch with satisfaction as the dust and hair fall away from the coat, leaving it slick and clean. Start with a fuzzy, muddy yak and finish with a gleaming steed.

The Daily Grooming Routine

The daily grooming session is an important ritual. Not only does it keep the horse clean, but it also has many other benefits: Bonding between owner and horse; ground-training and manners refreshers for the horse as needed; an opportunity to check the horse for physical problems such as injuries, swellings, skin problems, weight loss or gain, or behavioral

LEFT: Horses engage in mutual grooming to express and establish friendship.

changes; bringing circulation to the skin; lightly massaging the muscles; and distributing the skin's natural oils throughout the coat.

The basic routine is as follows:

1. Using a curry in one hand and a stiff-bristled body brush in the other, knock off any dried clumps of mud. Wet mud is more difficult to remove. It's best to either hose it off or wait for it to dry, and then brush it off. If the horse is shedding heavily in spring, use a shedding blade to remove as much loose hair as possible.

2. Use the curry firmly in small, circular motions over the muscular parts of the horse's body (neck, chest, shoulders, barrel, and haunches). The curry brings to the surface dirt, dry skin, and loose hairs that will later be brushed away.

3. Working from the front of the horse and moving toward the tail, use a medium-bristled body brush in the direction of hair growth to whisk away the debris loosened by the curry. Also use this brush on the legs and belly.

4. Switch to a soft-bristled brush to dust off any remaining debris left by the medium brush. Brush the horse's whole body in the direction of hair growth using short, brisk strokes.

5. Use a slightly damp washcloth or small towel to rub the horse's face and ears. Then use a small soft brush to set the hair as it dries.

6. Use a different damp cloth or sponge to clean under the horse's tail and between the hind legs as needed.

7. Pick the horse's hooves. Use a pick to remove packed-in dirt, carefully scraping it out from the clefts of the frog and checking for evidence of thrush (black, smelly material). Apply thrush treatment if needed.

8. Finger-comb the tail, removing any bits of hay or bedding. Do not brush or comb the tail unless it has just been washed and conditioned; otherwise you risk breaking the hairs.

LEFT Picking the front feet.

BELOW A soft brush provides the finishing touch to the coat.

RIGHT: Picking the hind feet.

LEFT: Hose off sweaty or muddy areas after each ride.

Comb the mane so that it all hangs on the same side of the neck, and comb the forelock to lie down flat.

9. For special occasions, use a rub rag to finish off the grooming job. Spritz a towel lightly with plain water or water mixed with a grooming product such as Show Sheen or baby oil. Rub the horse's body, removing all specks of dust and loose hair. Finish by using the soft brush to set the hair.

After your ride, brush the horse again with the soft brush to remove any arena dust. If he is just a little sweaty, use a medium-bristled brush on the sweaty areas (usually under the saddle and girth and on the neck) until the horse is dry. If he's very sweaty, hose or sponge off the sweat. After a hard ride on a hot day, a splash of liniment in the bath water is cool and refreshing and helps cut through the sweat. Rinse the neck, chest, saddle and girth area, belly, between the hind legs, and the head. (Be careful not to get any liniment in the horse's eyes.) Use a sweat scraper to remove as much water as possible, and towel-dry the lower legs to prevent fungus from developing.

The Quick Pre-Ride Grooming Session

You may not always have time for a full grooming session. If you're short on time before a ride, be sure to follow this routine to make sure the horse is comfortable during the ride:

*   Brush saddle and girth area with a hard brush.
*   Brush mud off legs, being especially careful if you'll be using wraps or boots.

- Pick hooves.
- Check bridle area and brush if needed.

Then just tack up and you're good to go. Be sure to do a thorough grooming the next day.

Treating Minor Injuries and Skin Conditions

In your daily grooming sessions, you will encounter the occasional minor laceration or skin irritation. Horses in group turnout situations use their teeth and hooves to communicate and to establish dominance, so it's only natural that a horse will come in with an occasional bite mark. (A horse who regularly gets bitten or kicked, however, should probably be moved to another pasture with more appropriate companions.) A small bite mark or abrasion is nothing to worry about, but treat it immediately to prevent infection or scarring.

Minor Cuts

Horses have quite thick skin, so many wounds simply result in the loss of the hair and top layer of skin. There's no bleeding, and the wound heals very quickly. For these types of injuries, simply clean the area with water and a soft cloth, and apply a thin layer of an ointment like Bag Balm, which moisturizes the tissue to promote healing and creates a barrier against dirt and bacteria.

A deeper wound with bleeding calls for cleaning with an antiseptic scrub such as Betadine. Dry the wound and apply a veterinary antibiotic ointment. Repeat this treatment daily until the wound heals. Wounds on the lower legs require extra attention, since they are closer to the ground and are exposed to bacteria in the soil and manure. A minor skin injury on a lower leg can easily become a serious infection. The fungal disease known as scratches can quickly take hold. Consider wrapping the injury with cotton batting and a standing wrap if the injury is large or actively bleeding. Call the vet if you aren't sure what to do or if the injury starts to look infected (inflamed, hot, and possibly oozing pus).

Note: A puncture wound involving a joint is a very serious injury. Call your vet immediately. If the puncture penetrates the joint capsule, severe infection can set in very quickly and can be life-threatening. Any injury to the eye or a puncture wound in the sole of the hoof (as from stepping on a nail) is also cause for an immediate call to the vet due to the severe risks associated with infection.

Skin Conditions

Horses are prone to a number of skin conditions, from allergic reactions to fungal infections. Some of the most common are hives, rain rot, and scratches (pastern dermatitis).

Hives are raised welts on the horse's skin, and are caused by an allergic reaction. They may be caused by a new type of feed, an insect, or something in the horse's environment. They may be itchy or not. For first aid, witch hazel or calamine lotion applied topically can help soothe the itch. In the long term, you'll need to identify the source of the allergy and remove it. For example, if it is insects, try using a different type of fly spray or a fly sheet, or perhaps even bring the horse into his stall at the time of day when the offending insects are feeding.

Rain rot is a fungal infection that typically affects the skin of the back, hindquarters, and neck and is commonly seen in neglected or malnourished horses. It presents as black or gray, scaly patches and hair loss on the affected areas. The direct cause is exposure to too much moisture (as from being left out in the rain) combined with too little grooming and inadequate nutrition to allow the body's immune system to function well. Clean thoroughly, treat with topical antifungal ointments, groom regularly, and protect the horse from rain while he heals. Review his diet and consider changing it, increasing it, or adding a vitamin/mineral supplement to make sure he is getting everything he needs.

Scratches is the common name of a fungal infection of the pasterns and fetlocks. You will notice raised bumps on the skin, developing into crusty scabs that are painful when picked at. Scratches can spread quickly, and left untreated can become quite severe, causing generalized inflammation of the lower limbs. It is caused by wet, unsanitary conditions, usually with horses standing in muddy, manure-filled paddocks, although even horses in drier conditions can get it. Some horses seem to be naturally more susceptible, and have to be treated preventatively to make sure they don't develop scratches each spring. For such a horse, keep the fetlock hair clipped short, make sure his living environment is dry and clean, and be sure to dry the lower legs any time they become wet.

To treat an existing case of scratches, bathe the affected area with an antiseptic scrub, dry thoroughly, and coat the scabby area in an ointment such as ichthammol, and leave it on overnight. This will penetrate and moisturize the scabs, allowing them to be picked off more easily. The next day, repeat the scrubbing, attempting to remove any scabs that will come off. Be careful, and be aware that the process may be painful for the horse. If some scabs are too painful to remove, leave them alone and repeat the ichthammol dressing and scrub-

RIGHT: The first step in treating any skin condition is usually to wash the area with an antibiotic or antifungal shampoo.

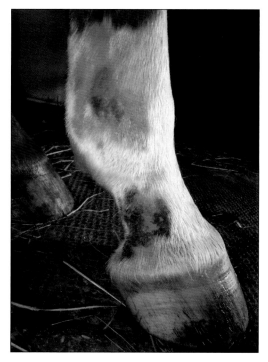

ABOVE: An especially nasty case of scratches.

ABOVE: To help prevent scratches from developing on a horse that is prone to the infection, towel-dry the legs thoroughly anytime they become wet.

bing cycle. Another option is to create a homemade remedy made from equal parts triple antibiotic ointment, zinc oxide diaper rash cream, and Monistat yeast infection treatment. Smear this paste onto the affected area twice daily. Before each application, gently wipe off any accumulated residue, but do not wash or scrub. Keep applying this treatment until all sign of infection is gone and new hair is growing back normally. It may take several days to clear up the infection. Be vigilant, and keep the horse in a dry area in between treatments. If scratches, or any other skin condition, does not begin to clear up after several days of treatment, call your vet for further advice.

Rain rot, scratches, and other fungal infections like thrush are signs that you may need to review your horsekeeping practices. Is your paddock too muddy? Do you pick out your run-in shed daily, or is it full of manure? Is there enough clean, dry bedding in your stalls? Do you groom your horses and pick their feet often enough? Perhaps most fundamentally, is your nutrition program up to par? If a horse is not getting enough to eat, or if his diet is lacking in the necessary vitamins and minerals, his immune system can suffer, allowing fungal infections to take hold. These types of infections can serve as the canary in the coal mine—letting you know that something is amiss before the problem becomes even more severe.

Handling and Riding

"Light off the leg, soft in the hand;
Ride the horse and not the plan.**"**
—Unknown

Handling Your Horse

Every time you ride or handle your horse, you are training him. He is learning either to respect you or not. Horses are extremely large animals with hard hooves and sharp teeth. If they learn to disrespect people, they can become unruly and dangerous. It's a vicious cycle—an owner lets her horse get away with some inappropriate behaviors, the horse starts testing the waters by threatening to bite or kick, the owner becomes frightened and backs off, and suddenly the horse knows he is bigger and stronger than you. This is not because horses are inherently malicious or because they want to bite and kick us. It's because, as herd animals, they need a leader. Most horses actually prefer to be the follower, but if the human doesn't step up and act as a leader, the horse assumes it's his job to become the leader of your little herd of two. This doesn't mean that horses need to be bullied and dominated into submission. Consistent, firm, yet gentle handling is most effective.

BELOW: Horses establish leadership using body language, such as this bite threat asking the less dominant horse to move away.

Expect your horse to be obedient and respectful toward all humans.

For a difficult horse or a novice owner who needs help learning about horse–human communication, clinics and training opportunities exist, especially within the field of natural horsemanship, which emphasizes handling and ground training (see below). However, opportunities for training your horse on the ground occur multiple times a day every day.

Expect and demand obedience when leading, grooming, tacking, tying, or working around your horse. Again, a gentle yet firm hand is best. Insist that your horse walk at your side when being led, rather than lagging behind or surging ahead. Insist that he stand quietly while being groomed. Insist that he pick up his feet promptly when you request it. Correct disobediences immediately to prevent further problems from developing. Most horses will be happy to know that they have such a strong and consistent leader keeping them safe and

Rules of Thumb for Horse Handling

- Never interact with a horse when you feel fear or rage. Take a step back, walk away, and calm down before moving on.
- When disciplining a horse, there must always be a "right answer" for him. There must be a correct way for him to behave that makes the discipline stop.
- Physical discipline is only appropriate when the horse threatens you first. Do not strike your horse for not standing still, for not being attentive enough, for not backing up quickly enough. A bite or a threat to bite may be met with a prompt, firm smack on the nose and a firm, "No!" A kick or threat to kick can be addressed with a verbal correction or growl and a smack of a crop on the offending leg. Don't let these behaviors go. They will only escalate.
- Discipline must occur within three seconds of the infraction for the horse to be able to make the association with his behavior.
- Always end a training session on a good note, whether riding or doing ground work. If you or the horse become frustrated or upset when trying to learn a new skill, go back to something easy and familiar before finishing your training session to boost your confidence again.

telling them what to do, so this kind of handling actually helps them relax. A few horses may try to "test" you more often, and may require an even firmer hand. Just as with riding, if you ever find that you're afraid of your horse, seek professional help from a trainer who can work with you and the horse to get you back on the right track. Fear is detrimental to your relationship with your horse, and will only cause his behavior to worsen when he sees that he's in control.

Safety Considerations

Wear a helmet. To keep yourself safe when out on the trail, the number one rule is to always wear a helmet. Accidents can happen even on the safest horse, and when outside the ring the chances that you might hit your head on a rock or tree are all too high. If the terrain is quite treacherous or you have reason to believe your horse might be spooky or hot, it's not a bad idea to wear a safety vest as well.

Carry a cell phone. Second, always carry a cell phone *attached to your body*, not to the saddle or the horse. If you fall or otherwise become separated from your horse, your phone won't do you much good if it's attached to the saddle and the horse is hightailing it back to the barn. Stash it in a zippered pocket or a holster attached to your arm, leg, or the back of your belt. (Your hip is not a great place for a phone either—if you fall, you're fairly likely to land on your hip, potentially breaking your phone.)

Wear bright colors. Bright, bold colors will make you more visible in the woods, alerting anyone else out there—hikers, mountain bikers, or hunters—to your presence

ABOVE: This rider is well outfitted in bright colors that make her highly visible to other trail users.

more quickly. The sooner they see you, the sooner they can leash their dogs, halt their bikes, or, in the case of a hunter, fail to shoot at you. In addition, if an accident were to occur, leaving you incapacitated or unconscious, bright clothing will make it easier for searchers to find you. Whenever you ride on the road, take care to make yourself as visible as possible with bright, light colors so drivers will easily see you and slow down. Reflective material is also beneficial in low-light conditions, such as early morning or evening.

Use the buddy system. Any time you ride out alone, make sure someone knows when you're leaving, when you expect to be back, and where you plan to ride. If you don't return at the expected time, they'll know where to go looking for you. Any time I ride alone (whether on the trail or in my field) I text my husband right before mounting. He knows to expect a follow-up text from me in about an hour to let him know I'm back. If he doesn't hear from me, he calls, and, although we've never had to use this plan, he would come looking for me or call emergency services if I don't answer.

RIGHT Any time you ride alone, make sure someone knows where you're going and when you'll be back.

Carry a first-aid kit. For short jaunts, this isn't really necessary, but on a long ride in a large wilderness area, it's wise to carry a small first-aid kit capable of treating human and equine injuries. It can fit in a fanny pack or saddle bag and should include VetWrap, gauze, antibiotic ointment, a sharp pocket knife or Swiss army knife, hoof pick, Band-Aids, epi pen, tweezers, hydrogen peroxide, antibiotic wipes, hand sanitizer, Tylenol, bottled water, and snacks such as granola bars or trail mix.

Every Time, Every Ride

There are currently no laws mandating that you must wear a riding helmet while mounted on a horse. But common sense and safety require that you do. Most English disciplines, particularly those that emphasize jumping, traditionally feature a helmet as part of regular riding attire. Dressage was historically an exception to that rule, with many riders schooling with no helmet and FEI riders wearing top hats at shows. However, recent high-profile head injuries, such as those of top dressage rider Courtney King-Dye, have started to change that trend. Now all riders through Fourth Level, as well as all riders under age eighteen regardless of level, at USEF-sanctioned dressage shows must wear helmets, and adult riders above Fourth Level have the option of the traditional top hat or protective headgear.

The vast majority of Western riders do not wear helmets. I'm not entirely sure why that is, but I believe it's mainly just tradition and trend. Recently, most governing bodies for horse shows have changed their rules, which formerly required Western attire including a cowboy hat, to allow riders who wish to do so to wear helmets. Hopefully we are starting to see a sea change in attitudes toward helmets among Western riders.

Still, you will often hear the argument that a rider does not need to wear a helmet because she rides in a Western saddle. I suppose that's because she can grab the horn in the event of a buck or a bolt. But no saddle will protect your head if your horse slips and falls under you, or rears and flips over backward. Many riders argue that their horses are so safe and quiet, there's no risk of a fall. But even under the best of circumstances, stuff happens. Any horse is capable of a spook, a misstep, or even a slip and fall, resulting in an unscheduled dismount for the rider. Above all, it is of the utmost importance that children be taught the value of a helmet and be required to wear one when mounted. You might as well set a good example and wear one yourself.

No one can force you to wear a helmet when riding your own horse on your own property, but there's really no compelling reason not to wear one. Maybe you'll feel foolish in front of your helmetless buddies. Maybe you're worried about messing up your hair. Maybe you enjoy the feel of the breeze across your scalp. None of these really holds a candle to the risk of concussion, traumatic brain injury, lifelong mental impairment, or death. Just wear your helmet—every time, every ride.

INDEX